CDM Regulations 2007
Procedures Manual

Acknowledgement

The author would like to acknowledge the patience and constructive dialogue continually provided by Julia Burden and the support of his wife, Linda, without whom this publication would never have been completed.

CDM Regulations 2007 Procedures Manual

Third Edition

Stuart D. Summerhayes
BSc, MSc, CEng, MICE, FaPS

Blackwell
Publishing

© 2008 by Blackwell Publishing Ltd

Blackwell Publishing editorial offices:
Blackwell Publishing Ltd, 9600 Garsington Road, Oxford OX4 2DQ, UK
Tel: +44 (0)1865 776868
Blackwell Publishing Inc., 350 Main Street, Malden, MA 02148-5020, USA
Tel: +1 781 388 8250
Blackwell Publishing Asia Pty Ltd, 550 Swanston Street, Carlton, Victoria 3053, Australia
Tel: +61 (0)3 8359 1011

First published 2008 by Blackwell Publishing Ltd

2 2009

ISBN: 978-1-4051-4002-7

Library of Congress Cataloging-in-Publication Data

Summerhayes, Stuart.
CDM regulations : 2007 procedures manual / Stuart D. Summerhayes. – 3rd ed.
p. cm.
Includes bibliographical references and index.
ISBN 978-1-4051-4002-7 (pbk. : alk. paper)
1. Construction
industry–Safety regulations–Great Britain. I. Title.
KD3172.C6S86 2008
343.41′078624–dc22
2007037122

A catalogue record for this title is available from the British Library

Set in 11/14pt Plantin by SNP Best-set Typesetter Ltd., Hong Kong
Printed and bound in Singapore by Fabulous Printers Pte Ltd

The publisher's policy is to use permanent paper from mills that operate a sustainable forestry policy, and which has been manufactured from pulp processed using acid-free and elementary chlorine-free practices. Furthermore, the publisher ensures that the text paper and cover board used have met acceptable environmental accreditation standards.

For further information on Blackwell Publishing, visit our website:
www.blackwellpublishing.com

Contents

HOW THIS MANUAL WORKS

This procedural manual is divided into sections to cover the full remit of obligations imposed on the:

- client
- designer(s)
- contractor(s)
- CDM co-ordinator
- principal contractor.

Each of these functions is qualified by a flowchart and detailed checklist. The flowchart provides a chronological route through the Regulations for each duty. Key to this route is the node box, which identifies obligatory duties and procedural options together with reference to the corresponding Regulation and further description where appropriate.

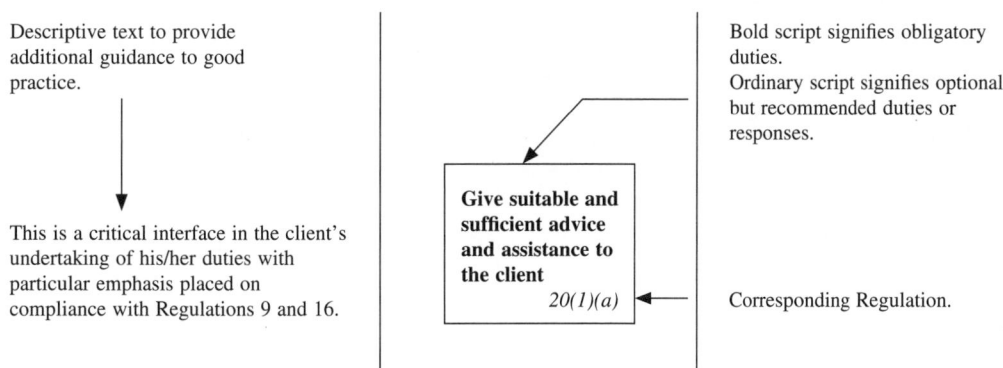

The flowchart provides an illustrated link between successive duties.

The checklist documents accountability and provides a detailed procedural step-by-step approach in fulfilment of obligations. It serves as the audit trail through the project, signifying what action needs to be taken and providing a record of what action has been taken, endorsed by the signature of the responsible person.

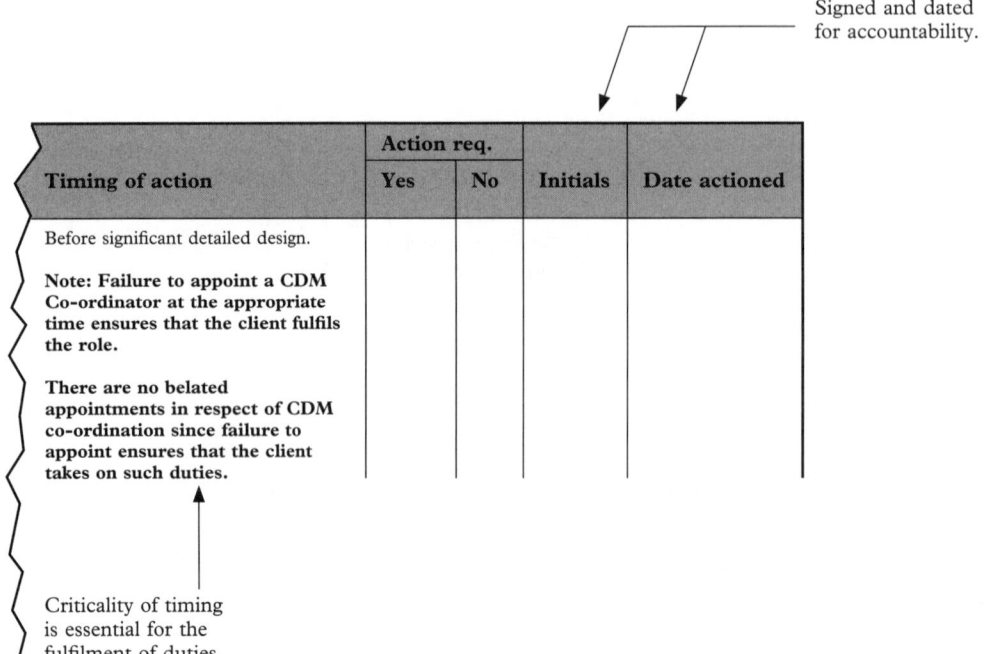

CONTRACT			
Reg.	**Stage**	**Procedure**	**Description**
14(5)	Post initial design stage Before the start of significant detailed design	**NOTIFICATION OF APPOINTMENT**	**The appointment of the CDM co-ordinator must be made in writing by the client.** It would be imprudent to fulfil this role without the written confirmation of appointment. Where the role of CDM co-ordinator has been fulfilled within the client organisation either by design and/or default care needs to be exercised and conditions/qualifications made about the service being subsequently provided.

Corresponding
Regulation

Bold text signifies obligatory duties.
Ordinary script signifies optional but recommended duties or responses.

Descriptive text provides additional guidance for good practice.

Signed and dated for accountability.

	Action req.			
Timing of action	**Yes**	**No**	**Initials**	**Date actioned**
Before significant detailed design. **Note: Failure to appoint a CDM Co-ordinator at the appropriate time ensures that the client fulfils the role.** **There are no belated appointments in respect of CDM co-ordination since failure to appoint ensures that the client takes on such duties.**				

Criticality of timing is essential for the fulfilment of duties.

An example of a Pre-Construction Information – Tender Stage Document and Project (Health and Safety) Risk Register is included together with checklists and agendas as aide-memoires in support of the management of the process.

Section 1
INTRODUCTION

The Construction (Design and Management) Regulations (CDM Regulations) 1994 were the United Kingdom's response to transpose the Council of the European Communities Directive 92/57 EEC entitled 'The Minimum Health and Safety Requirements at Temporary or Mobile Construction Sites' into British law. They arose from member consensus that the entire responsibility and management of health and safety throughout the construction process was both fragmented and unco-ordinated.

These regulations have now been revised in the form of the CDM Regulations 2007, which were laid before Parliament on the 15 February 2007 as Statutory Instrument 2007 No. 320, and were enacted on 6 April 2007. These regulations apply to all construction work in Great Britain and its territorial waters (12 miles offshore).

The revision has allowed greater alignment with the existing Directive, which has not changed, and has built on the industry's experience of working with the original regulations. These experiences relate to both the construction industry's cultural response and shortcomings identified in the interpretation and enactment of the original regulations.

Criticism attached to the original regulations related to:

- excessive bureaucracy
- belated appointments
- baseline not benchmark approaches
- enforcement difficulties
- reactive not proactive responses.

The CDM Regulations 2007 are accompanied by an Approved Code of Practice (ACoP) (L144), which is to be supported by sectorial guidance provided by industry in due course.

The new regulations emphasise the project management approach and are process related with **co-operation**, **co-ordination** and **communication** fundamental to the delivery of effective health and safety management within the holistic concept of the **integrated team**.

The Health and Safety Commission, in collaboration with a Construction Industry Advisory Committee working group, prepared the documents (the regulations and ACoP), which are written in a more explicit language form directed towards:

- simplification
- maximising flexibility
- planning and management of risk
- improved competence of duty holders
- improvements in process and practice
- integration via co-operation and co-ordination
- providing the right information to the right people at the right time.

Major changes have been made with the introduction of the **CDM co-ordinator** and the concept of information flow via **pre-construction information**. Greater liability for all duty holders is evident and due emphasis placed on professional competence as outlined in Appendices 4 and 5 of the corresponding ACoP (L144) and a centre of gravity shift towards the client, who must exercise more control over the entire process.

The CDM Regulations 2007, unlike their predecessors, combine two major items of construction-related legislation in one coherent document, so the new regulations are both an update of the CDM Regulations 1994 (plus 2000 amendment) and the remnants of the Construction (Health, Safety and Welfare) Regulations 1996 (less the content of the Work at Height Regulations 2005). This will require familiarisation since the new Regulations are now logically structured as:

Part 1	Regulations 1 to 3	**INTRODUCTION**
Part 2	Regulations 4 to 13	**GENERAL MANAGEMENT DUTIES: ALL CONSTRUCTION PROJECTS**
Part 3	Regulations 14 to 24	**ADDITIONAL DUTIES WHERE PROJECT IS NOTIFIABLE**
Part 4	Regulations 25 to 44	**DUTIES RELATING TO HEALTH AND SAFETY ON CONSTRUCTION SITES**
Part 5	Regulations 45 to 48	**GENERAL**

The CDM Regulations 2007 therefore become the major piece of legislation within the portfolio of construction-related legislation and seek to influence the following industry statistics:

- an unprecedented 30% increase in fatalities over 2005–2006
- 3677 major injuries to employees (2005–2006)
- 7492 over-3-day injuries to employees
- 86 000 employees suffering from work-related ill-health
- 3.2 million working days lost per year (injury and ill-health).

However, any improvements to these statistics will require not only legislative change but also an industry response in moving attitudes and ownership closer to the integrated team concept identified in the Egan Report *Rethinking Construction*.

The cost benefits of effectively managed projects have been well publicised by the Health and Safety Executive and Constructing Excellence, with due emphasis on the role of effective health and safety management. It should not be difficult for the construction industry to appreciate the valid argument surrounding commercial viability in its cultural move to embrace and move beyond basic compliance with criminal legislation as represented by the CDM Regulations 2007.

Much has been achieved in raising the awareness of health and safety issues over the last 12 years, with much more still to be done. This cannot be achieved by legislation alone and whilst the Regulations seek to build on the platform first established by the Health and Safety at Work, etc. Act 1974, with its emphasis on ownership through dialogue and consultation rather than imposition, industry culture must also shoulder its own responsibility.

As with project management generally it is communication failure that precedes project failure and all duty holders within the construction industry need to ensure that projects start off correctly in order to deliver successful health and safety management outcomes. This is also one of the objectives of the legislation.

All construction professionals, therefore, need management systems that are compatible and supportive of core business delivery. Indeed, there is much to be done before a proactive health and safety construction culture contributes

effectively to all stages of a project via its acknowledgement as an integral part of successful project management.

It should be noted that no legislation achieves the aspirations of all duty holders for it should be acknowledged that whatever the modifications and amendments embraced in the drafting of the new regulations, it is the same construction industry to which these regulations are directed.

For these regulations to work and achieve the truly integrated holistic team-based approach that moves interminably towards a benchmark standard, industry must discard some of its practices and seek to embrace much of the ethos and philosophy first laid out in the Egan report.

The reader should appreciate that the CDM model is closely aligned with the concept of the general project management model and is indeed an integral subset of project management. Both will continue to remain journeys of constant improvement for the intent is not to arrive but to constantly build on the experience of the current project. Lessons learnt from one project need to be taken into the objectives of the next project, with each team member striving to optimise the learning experience, thus avoiding a reinvention of the wheel scenario.

Over the 12 years' experience gained under the regime associated with the CDM Regulations 1994 it remained a disappointment that too many organisations were still looking to comply. It is a useful focus to remind ourselves that compliance is the baseline in that no duty holder can do less and stay above the law. However, there is nothing preventing that duty holder doing more in the furtherance of the best practice approach.

Accompanying the CDM Regulations 2007 is the clear message that the corresponding philosophy seeks to *manage risk and not the paperwork*. The effectiveness of health and safety management is not improved by the extent of the paperwork associated with the audit trail of documentary evidence. Key messages are capable of being lost because of the information overload associated with irrelevant information. All sectors of our industry must understand the implications of duty holder discharge within the conceptual embrace of competence and the comfort of proportionality.

Additionally, the duty holder must remain empowered to provide evidence of the process without being hampered by visibility of detail at each stage. Each duty holder must be able to demonstrate, through the relevancy of focused documentary evidence, his effective contribution to health and safety management within the construction environment without being held to account by the minutiae of irrelevant detail.

Duty holders have always been charged with being contributors to health and safety management as well as communicators of information and outcomes integral to the process. The former requires awareness and competence, whilst the latter requires the usual communication skills of simplicity, succinctness and focus. None of this can be delivered without the organisational structure that supports team delivery.

These regulations are not about compliance – they are about improving and controlling the construction environment through compliance and beyond. This cannot be achieved without the cultural embrace of the industry to which they relate.

These regulations provide the industry with an opportunity to advance construction excellence and further promote team integration for the benefit of health and safety management.

Section 2
APPLICATION AND INTERPRETATION

2.1 Application

Regulation 3 Application

'1) These Regulations shall apply
    a) in Great Britain; and
    b) outside Great Britain as sections 1–59 and 80 to 82 of the 1974 Act by virtue of article 8(1)(a) of the Health and Safety at Work, etc. Act 1972 (Application outside Great Britain) Order 2001
2) Subject to the following paragraphs of this Regulation, these Regulations shall apply to and in relation to construction work.
3) The duties under Part 3 shall apply only where a project
    a) is notifiable; and
    b) is carried out for or on behalf of, or by, a client.
4) Part 4 shall apply only in relation to a construction site.
5) Regulations 9(1)(b), 13(7), 22(1)(c), and Schedule 2 shall apply only in relation to persons at work carrying out construction work.'

Thus:

- Parts 1 and 2   apply to all construction projects
- Part 3   applies only to notifiable projects
- Part 4   applies only in relation to the management of a construction site
- Part 5   general.

Basically there are two situations to consider and all construction projects regardless of complexity will fall into one of the following two categories:

(1) **non-notifiable** projects
(2) **notifiable** projects.

As noted in Regulation 2, a project is notifiable if the construction phase is likely to involve more than:

- 30 days, or
- 500 person days
of construction work.

Thus, unlike the CDM Regulations 1994, there are no special cases associated with:

- demolition/dismantling, or
- the five-person rule.

In respect of the former, all demolition/dismantling is now governed by the requirements of the following:

Regulation 29(2) Demolition or dismantling

'1) The demolition or dismantling of a structure, or part of a structure, shall be planned and carried out in such a manner as to prevent danger, or where it is not practicable to prevent it, to reduce danger to as low a level as is reasonably practicable.

2)   *The arrangements for carrying out such demolition or dismantling shall be recorded in writing before the demolition or dismantling begins.'*

Much demolition/dismantling itself is associated with the enabling works of the larger project and whilst all demolition must comply with Part 4 requirements, and particularly Regulation 29(2), it is the project itself that dictates whether or not it is notifiable or non-notifiable in status.

Additionally, relevant parties must be aware of:

Regulation 25(2)

'Every person (other than a contractor carrying out construction work) who controls the way in which any construction work is carried out by a person at work shall comply with the requirements of Regulations 26 to 44 insofar as they relate to matters which are within his control.'

The above has an impact on all those who control the way in which work has to be undertaken and is focused on *'every person who controls the way'* rather than any particular duty holder.

Table 2.1 outlines the main responsibilities of duty holders.

Table 2.1
THE MAIN RESPONSIBILITIES OF DUTY HOLDERS

| DUTY HOLDER | Section | Part 2 | | | | | | | | | | Part 3 | | | | | | | | | | | Part 4 | |
	Regulation	4	5	6	7	8	9	10	11	12	13	14	15	16	17	18	19	20	21	22	23	24	25(2)	26 to 44
Client		X	X	X	X	X	X	X		X		X	X	X	X								X	(X)*
Designer		X	X	X	X				X	X						X	X						X	(X)*
Contractor		X	X	X	X					X	X												X	X
CDM co-ordinator		X	X	X	X													X	X				X	(X)*
Principal contractor		X	X	X	X					X	X						X			X	X	X	X	X

Note: CDM co-ordinators and principal contractors are only appointed on notifiable projects. *Applies only if Regulation 25(2) applies.
Unshaded Regulations apply to all construction projects. Shaded regulations additionally apply to notifiable projects.
Part 1 (Regulations 1 to 3) deals with interpretations and application.
Part 5 (Regulations 45 to 48) deals with general, civil liability, enforcement in respect of fire, transitional provisions, and revocations and amendments.

2.2 Interpretation

Construction work means the carrying out of any building, civil engineering or engineering construction work and includes:

Regulation 2(1)

'(a) the construction, alteration, conversion, fitting out, commissioning, renovation, repair, upkeep, redecoration or other maintenance (including cleaning which involves the use of water or an abrasive at high pressure or the use of corrosive or toxic substances), de-commissioning, demolition or dismantling of a structure,

(b) the preparation for an intended structure, including site clearance, exploration, investigation (but not site survey) and excavation, and the clearance or preparation of the site or structure for use or occupation at its conclusion,

(c) the assembly of prefabricated elements to form a structure or the disassembly on site of prefabricated elements which, immediately before such disassembly, formed a structure,

(d) the removal of a structure or of any product or waste resulting from demolition or dismantling of a structure or from disassembly of prefabricated elements which, immediately before such disassembly, formed a structure, and

(e) the installation, commissioning, maintenance, repair or removal of mechanical, electrical, gas, compressed air, hydraulic, telecommunications, computer or similar services which are normally fixed within or to a structure but does not include the exploration for an extraction of mineral resources or activities preparatory thereto carried out at a place where such exploration or extraction is carried out.

but does not include the exploration for, or extraction of, mineral resources or activities preparatory thereto carried out at a place where such exploration or extraction is carried out.'

Structure means:

'(a) any building, timber, masonry, metal or reinforced concrete structure, railway line or siding, tramway line, dock, harbour, inland navigation, tunnel, shaft, bridge, viaduct, waterworks, reservoir, pipe or pipeline, cable, aqueduct, sewer, sewage works, gasholder, road, airfield, sea defence works, river works, drainage works, earthworks, lagoon, dam, wall, caisson, mast, tower, pylon, underground tank, earth retaining structure, or structure designed to preserve or alter any natural feature, fixed plant and any other structure similar to the foregoing or,

(b) any formwork, falsework, scaffold or other structure designed or used to provide support or means of access during construction work
and any reference to a structure includes a part of a structure.'

2.3 Additional exclusions

Note the following additional exclusions to the construction work definition as identified in paragraph 13 of the ACoP:

- *Putting up and removal of marquees and tents.*
- *General maintenance of fixed plant other than major refurbishment and alterations of structures in their own right.*

- *Tree planting and general horticultural work (as distinct from similar work associated with larger remediation schemes).*
- *Positioning and removal of lightweight partitions – mention here is made of work associated with exhibition stands, although some exhibition models move into structural assemblies and interpretations should not be taken too literally.*
- *Surveying as distinct from invasive investigation work.*
- *Work in conjunction with vessels such as ships and mobile offshore platforms.*
- *Off-site manufacture.*
- *Fabrication of elements which will form parts of offshore installations.*
- *The construction of fixed offshore oil and gas installations at the place where they will be used.*

Thus, with the exceptions above, the distinction between the survey and investigation, and the exploration and investigation of mineral resources (open-cast mining) the CDM Regulations 2007 apply to all construction work.

Care should be exercised that site clearance, site investigate and site preparation works should be seen as part of the project regardless of the period between such weeks and the main construction works themselves. Thus, for a notifiable project such early works will themselves require the appointment of CDM co-ordinator and principal contractor and appropriate documentation and control.

Section 3
ALL DUTY HOLDERS

3.1 Introduction

Parts 1 and 2 of the regulations apply to all duty holders and seek to emphasise the rudiments of the project management approach, namely:

- Competence   Regulation 4
- Co-operation   Regulation 5
- Co-ordination   Regulation 6
- Communication   Regulations 10(2), 11(6), 13, 18, 19, 20, 21, 22, 23

together with the basic tenet of health and safety management enshrined in the General Principles of Prevention as outlined in Schedule 1 of the Management of Health and Safety at Work Regulations 1999 and now endorsed by Regulation 7.

The project management model to which the CDM model aspires is based on the need to start every project with adequate preparation and control from the twin perspectives of foreseeability and proactivity, and throughout the project to continuously monitor and review the effectiveness of these controls to ensure the delivery and management of safe and suitable systems of work.

This can only be effectively achieved through the integration offered by the holistic process, where ownership of health and safety management is delivered by the team. Fragmented teams cannot deliver the success demanded and undermine the very concept of collective ownership of the health and safety issues.

The integrated team is therefore an essential prerequisite on all projects in the delivery of effective project management as well as effective health and safety management. Fragmentation undermines the basics of the team approach within the holistic concept and therefore contributes to communication failure and lack of control. This is unacceptable on professional, moral and legislative grounds.

Project management integration also embraces other risk management facets, including financial management, quality management, time management, and health and safety management. All are inextricably linked, with failure in one inevitably leading to failure in others.

However, it is health and safety management statistics that advertise the dire consequences of control failure, with many of the effects still hidden and not yet confronted.

As such the project management words shown in Table 3.1 resonate throughout the regulations to remind all duty holders of the need for a proactive, ongoing, team-based approach.

Table 3.1

PROJECT MANAGEMENT DESCRIPTORS FOUND IN THE REGULATIONS

Action required	Duty holder	Regulation
'maintained and reviewed'	Client	9(2)
'plan, manage and monitor'	Contractors	13(2)
'shall ensure . . . are complied with throughout the construction phase'	Contractors	13(7)
'revised as often as may be appropriate'	Client	17(3)(b)
'might justify a review of the construction phase plan'	Contractor	19(2)(a)
'take appropriate action . . . where it is not possible to comply'	Contractor	19(2)(b)
'planned, managed and monitored'	Principal contractor	23(1)(a)
'made and implemented'	CDM co-ordinator	20(1)(b)
'review and update'	CDM co-ordinator	20(2)(e)
'consult a contractor before finalising'	Principal contractor	22(1)(g)
'plan, manage and monitor the construction phase'	Principal contractor	22(1)(a)
'review, revise and refine'	Principal contractor	23(1)(b)
'consult those workers or their representatives in good time'	Principal contractor	24(b)

Figure 3.1
HOLISTIC DIAGRAM

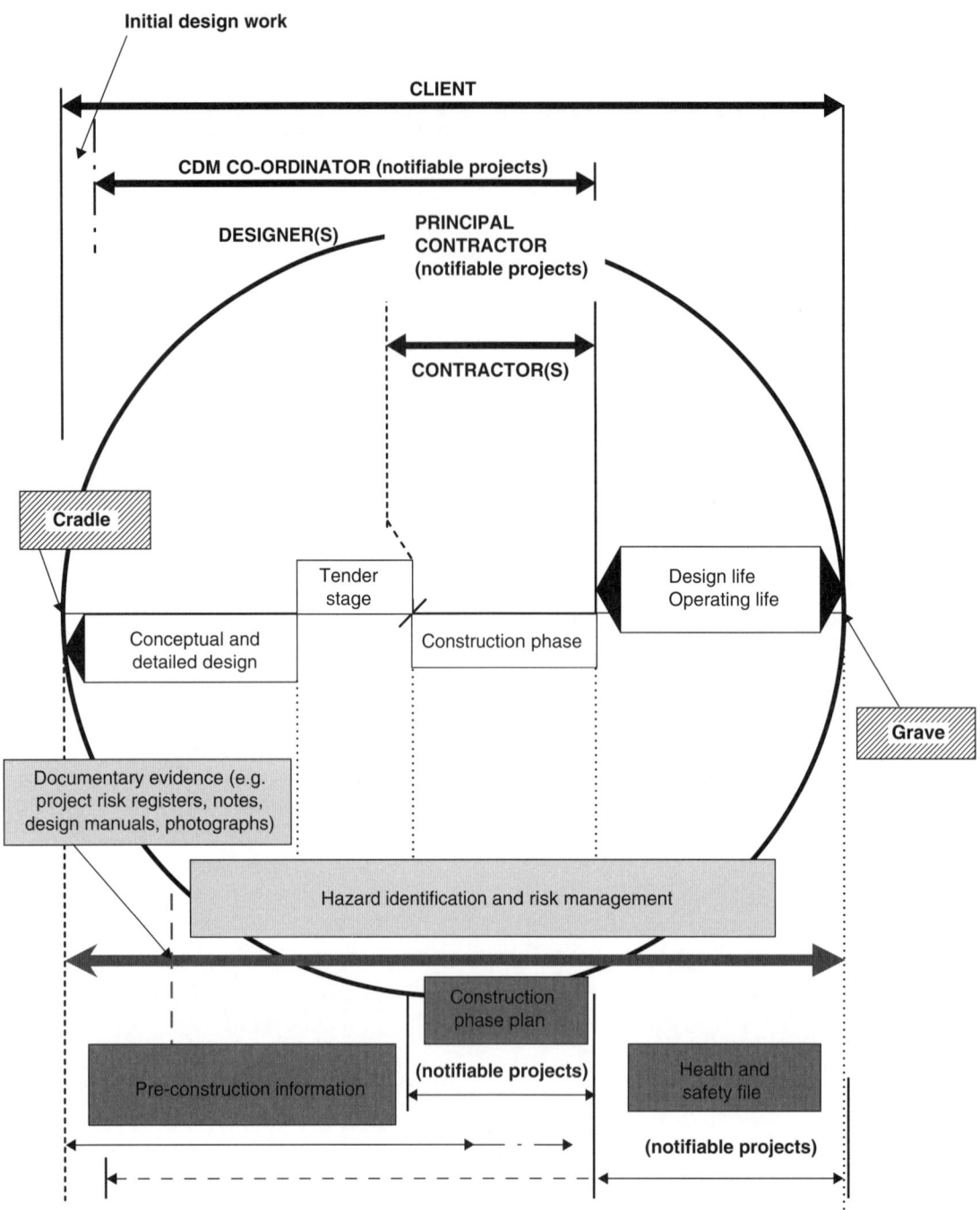

Figure 3.1 Holistic diagram

In respect of the holistic diagram in Figure 3.1 the following should be noted:

All projects:

- All duty holders have implicit health and safety duties once they enter the holistic circle, i.e. immediately they become involved in a project.
- Figure 3.1 is based on a conventional procurement strategy. For different strategies there will be greater overlaps between design and construction.
- For the design process, the emphasis should be on hazard identification and risk management strategy rather than design risk assessment. The latter is simply a facet of the overall strategy.
- Hazard identification and risk management strategy need to embrace all the foreseeable issues arising out of design itself and must include issues of constructability, usability, maintainability, serviceability and removability as well as replacement.
- Pre-construction information is an information flow that continues up until the construction activity itself takes place. Hence pre-construction information could still be delivered during the construction phase. This is required on all projects and emanates from the client (Regulation 10).

Notifiable projects only:

- The appointments of the CDM co-ordinator and principal contractor only occur on the notifiable project.
- The appointment of a CDM co-ordinator should be made as soon as practicable after the initial design stage. Note the ACoP (paragraph 66) distinction between *'initial design work'* and *'significant detailed design'*.
- The construction phase plan is the document that articulates the safe and suitable systems of control to be exercised by the principal contractor throughout all the construction stages of a project. It is only required on a notifiable project, but appropriate information would be required on all projects, e.g. workplace risk assessments and methodologies, etc. It is critical that this key document is sufficiently developed before any construction work starts.
- The management of the health and safety file begins simultaneously with the appointment of the CDM co-ordinator and control over this management process must be exercised right throughout the project by the CDM co-ordinator.

In Figure 3.1 the circle represents the integrated holistic team-based approach and is a reminder that fragmented teams rarely deliver successful outcomes and never deliver a fully successful outcome.

Section 4
THE CLIENT

The client is seen as instrumental to the entire strategy to which the regulations relate, since he remains best placed to influence the whole process. The client is at the head of the supply chain and exercises control over the project in terms of contract arrangements and management of funds, and is therefore the key duty holder who sets the whole tenure of the project.

The identity of the client, as with other duty holders, needs to be established as early as possible, particularly since the client or their representative must sign the notification that goes off to the Health and Safety Executive (HSE) or to the Office of Rail Regulation on a notifiable project. The identification can become complicated and reflective of procurement strategy but can be facilitated by considering which entity:

- heads the procurement chain
- enters into the contract
- manages the funding arrangements
- makes the strategic decisions.

The client means:

'*a person who in the course or furtherance of a business*

   a)   *seeks or accepts the services of another which may be used in the carrying out of a project for him; or*
   b)   *carries out a project himself.*'

Domestic clients generally are exempt from the CDM Regulations 2007, unless Regulation 25(2) applies since this has implications for all those who control the way in which construction work has to be undertaken.

It should be noted that the CDM Regulations 2007 do not allow the client to appoint a client's agent, as did the CDM Regulations 1994. However, such an appointment already made under the CDM Regulations 1994 can continue to function until five years (fixed-term maintenance contract) from the introduction of the CDM Regulations 2007 or until the project concludes for which the appointment was made, if sooner.

Regulation 25(2)   Application of Regulations 26 to 44

   '(2)   *Every person (other than a contractor carrying out construction work) who controls the way in which any construction work is carried out by a person at work shall comply with the requirements of Regulations 26 to 44 insofar as they relate to matters which are within his control.*'

This regulation could implicate all those who impose constraints on working practices via:

- access restrictions
- construction work alongside operational needs
- limitations through imposed procedures
- stipulated methodologies.

A distinction should also be made between the domestic client and domestic premises. The latter ensures duties are placed on others such as insurance companies, housing associations, developers and warranty providers, all of whom could function as clients under certain circumstances.

There are duties to be discharged by the client in respect of:

(1) All construction projects

Regulation 4	Competence
Regulation 5	Co-operation
Regulation 6	Co-ordination
Regulation 7	Principles of prevention
Regulation 8	Election by clients
Regulation 9	Client duty re managing projects
Regulation 10	Client duty re information
Regulation 12	Designs prepared outside Great Britain
(Regulation 25	Health and safety on construction sites)

and

(2) Notifiable projects

Regulation 14	Appointments where notifiable
Regulation 15	Client duty re information
Regulation 16	Start of construction
Regulation 17	Health and safety file

Clients have far more duties than others in accordance with their influential position in the supply chain. Not only is the client the major party that enters into the contract but the client also establishes the parameters of delivery that bind other duty holders into the holistic envelope that represents construction.

In embracing the changes associated with the CDM Regulations 2007 it is the client, of all the five duty holder groups, who must make the biggest cultural step in rising to the additional demands associated with liability and discharge of duties.

The client should note that duties to Regulations 4, 5, 6, 7, 8, 9, 10, 12 and 25(2) remain to be discharged on all projects. It is obvious that all clients, but particularly the lay client, need advice and assistance, not least because some smaller projects represent complex or high-risk construction work.

Such advice and assistance could be provided by:

• the professional team
• a competent person appointed under Regulation 7 of the Management of Health and Safety at Work Regulations 1999
• a CDM advisor (someone who would normally function as a CDM co-ordinator on a notifiable project).

Such high-risk projects are identified in the ACoP (clause 21) as:

• structural alterations
• deep excavations and those in contaminated ground
• unusual working methods or safeguards
• ionising radiation or other significant health hazards

- nearby voltage powerlines
- a risk of falling into water which is, or may become, fast flowing
- diving
- explosives
- heavy or complex lifting operations.

The ACoP reminds us that for the above and more complex projects generally, a more rigorous approach to co-ordination, co-operation and planning will be needed. Management arrangements would need to respond accordingly.

For notifiable projects the client can rely directly on the appointed CDM co-ordinator, but similar duties of health and safety management effectiveness must also be delivered by duty holders on non-notifiable projects. Whilst the effective discharge of all regulations is essential, the client needs in particular to focus attention on Regulations 9 and 10 for all projects and Regulations 14 and 16 for notifiable projects.

Regulation 9 requires the client to ensure management arrangements are established by all duty holders for the delivery of effective health and safety management and that such arrangements are monitored throughout the project to ensure that they remain effective. As such the client's team must ask the right questions at the right time. This duty does not translate into undertaking site audits per se because, as the ACoP advises, the client needs to seek confirmation that arrangements remain effectively in place. This can be achieved through carefully scripted agenda items.

Extra diligence needs to accompany the start of all projects to ensure that systems are established to:

- protect the workforce and the public
- satisfy the requirements for co-operation and co-ordination
- provide welfare provision
- secure the site.

These systems must remain in place throughout all projects.

Regulation 10 relates to the provision of pre-construction information by the client to all designers and those contractors who have been, or who might be, appointed by the client. This is the information stream that gathers capacity and momentum compatible with the iterative nature of project development. Thus information is directed from the stream to the recipients to include:

- relevant site information from historical sources
- the use of the building as a place of work
- the minimum period for planning and managing before construction work starts (the mobilisation period).

Such pre-construction information must be qualified and managed in its delivery to those receiving it to enable them to fulfil their statutory duties.

Regulation 14 requires clients on a notifiable project to appoint both a CDM co-ordinator and a principal contractor. The former must be appointed as soon as practicable after the initial design phase (Paragraphs 66 and 86 distinguish between '*initial* *design*' and '*significant detailed design*'), with encouragement to appoint the principal contractor as soon as procurement allows. The critical factor for the client to note is that failure to appoint a principal contractor at the correct time ensures that the client himself takes on those duties by default.

Regulation 16 represents primary control for the client in that the gateway represented by the start of the construction phase can only be accessed through the provision of a construction phase plan, which is under the ownership of the principal contractor. This must be sufficient in detail and content to demonstrate that suitable and sufficient systems of work have been developed and articulated to allow the early stages of construction to commence under suitably controlled conditions. The client exercises sanction over this start and must be satisfied with the sufficiency of this document before any work commences on site.

4.1
CLIENT FLOWCHART:
ALL PROJECTS

Initial design phase	Significant detailed design phase	Pre-construction phase	Construction phase	Post-construction phase
Appoint competent duty holders *4(1)*	**Appoint competent duty holders** *4(1)*	**Appoint competent duty holders** *4(1)*	**Appoint competent duty holders** *4(1)*	All duty holders whenever appointed must be assessed for competence. No person appointing can rely on self-certification arising out of Regulation 4(1)(b).
Co-operate and seek co-operation *5(1)*	**Co-operate and seek co-operation** *5(1)*	**Co-operate and seek co-operation** *5(1)*	**Co-operate and seek co-operation** *5(1)*	All duty holders must co-operate with and actively seek the co-operation of those involved in construction on the same or adjoining sites. **Note:** The project site is also an adjoining site to those around. This will remain a dynamic situation throughout the project.
Co-ordination *6*	**Co-ordination** *6*	**Co-ordination** *6*	**Co-ordination** *6*	Particularly relevant within any operational/occupied situation. Co-ordination must embrace the health and safety needs of those carrying out or affected by construction work.
General principles of prevention *7*				All duty holders must ensure that these are accounted for in the design, planning and preparation of the project.
Election by clients *8*	Where there is a group of clients involved in a project, one or more of them can elect to represent the group as the client. Such election must be in writing.	Nonetheless, certain duties remain to be discharged by all the parties, such as co-operation and provision of relevant information.		
Management infrastructure *9*	**Management infrastructure** *9*	**Management infrastructure** *9*	**Management infrastructure** *9*	This is an over-arching duty and such arrangements must be maintained and reviewed throughout the project. It has impact on all stages of the project and requires verification that others are discharging their duties competently from start to finish of the project. Thus roles and responsibilities must be established as well as effective lines of communication.
Pre-construction information *10*	**Pre-construction information** *10*	**Pre-construction information** *10*	**Pre-construction information** *10*	Such information is required to be passed to all designers and contractors who have been or may be appointed by the client. The flow of relevant information is not limited to the pre-construction phase.
Designs prepared or modified outside Great Britain *12*	Duties could fall to the client to ensure that all designers discharge duties compliant with Regulation 11 if the commissioning link is not established.			Such information includes: • information about the site • use of the structure • mobilisation period • existing health and safety file information.
		Even on non-notifiable projects the client's systems need to ensure modifications and amendments to existing health and safety files occur as a result of relevant work undertaken.	Existing health and safety file	

4.2
CLIENT FLOWCHART: NOTIFIABLE PROJECTS, ADDITIONAL DUTIES

Initial design phase	Significant detailed design phase	Pre-construction phase	Construction phase	Post-construction phase
Appoint CDM co-ordinator 14(1)	CDM co-ordinator must be appointed as soon as practicable after initial design.	**Note:** Appointment must be in writing		
Sign copy of initial F10 21(3)	The client or someone on his behalf must sign that the client is aware of his duties under these regulations.	If sent by electronic means client must ensure that he has approved the use of the electronic signature.		
Provide pre-construction information to CDM co-ordinator 15(a) and (b)	Client must promptly provide pre-construction information to the CDM co-ordinator, who must then ensure it goes out to all designers and contractors who have been or who may be appointed by the client.			
	Note: Appointment must be in writing	**Appoint Principal Contractor** 14(2)	This appointment takes place after the appointment of the CDM co-ordinator and must occur before the start of any construction work.	The earlier the appointment subject to the procurement strategy the better since the constructability perspective is enhanced.
		Control the start of construction 16(a) and (b)	No construction starts until the client is satisfied that the construction phase plan is sufficient and compatible with the requirements of Regulations 23(1)(a), 23(2) and 22(1)(c).	This sufficiency relates to safe and suitable systems of control in respect of early phase activities and the provision of adequate welfare facilities.
Provide relevant health and safety file information 17(1)	This is information likely to be needed for inclusion in the health and safety file. It relates to information in the client's possession or which is reasonably obtainable.		This must be handed over by the CDM co-ordinator. A relevant document must accompany partial handover.	Receive health and safety file from the CDM co-ordinator 20(2)(e)
			The client has ongoing duties after the handover of the project in respect of : ● ensuring the availability of the health and safety file to those who need access to it whilst at the same time maintaining possession of it ● maintaining the currency of the information in the health and safety file with relevant updates, etc. ● delivering the health and safety file to any new owner; change of ownership requires transfer of an up-to-date health and safety file ● an accompanying letter of explanation.	Make health and safety file available for inspection 17(3)(a) Revise health and safety file 17(3)(b) Deliver health and safety file to new owner 17(4) Ensure new owner is aware of nature and purpose of health and safety file 17(4)

4.3
CLIENT CHECKLIST

CLIENT CHECKLIST (Sheet 1 of 2)

The following procedures are requirements for all construction projects.

Contract No.	Reg.	Stage	Procedure	Description	Timing of action	Action req. Yes	Action req. No	Initials
1	8	Pre-initial design	**ELECTION BY CLIENTS**	A group of clients can elect one of them to be treated as the client. Such an election must be made in writing. Irrespective of such an election ongoing duties of co-operation and the provision of relevant information continue to apply to all.	Before the start of the project.			
2	4(1)(a)	Pre-appointment	**APPOINT COMPETENT DUTY HOLDERS**	All duty holders, i.e. designer(s) and contractor(s) appointed by the client must be assessed as competent. The client must be satisfied that the person/organisation appointed has the necessary competence to perform the function. The client cannot rely on any party self-certifying themselves as competent. The rigour of the assessment must relate to the complexity of the project.	Such assessment must be undertaken before appointment and must have appropriate rigour in the process of competence assessment.			
3		Pre-appointment	**IDENTIFY DUTY HOLDERS**	To fulfil the above duty the client must identify all designers and contractors who have been or who may be appointed by him.	Throughout the project.			
4	5	All stages	**CO-OPERATION OF DUTY HOLDERS**	The client and all other duty holders must seek the co-operation of those involved in construction not only on the project but also on adjoining sites. **Note:** Your project is itself an adjoining site.	Particularly at the start of the project but actively throughout the project period. This is a project dynamic and must be continuously fulfilled.			
5		All stages	**IDENTIFICATION**	There is a continuing need to identify all those with whom the client needs to co-operate. This could well relate to overlapping projects, etc.	Particularly at the start of the project but actively throughout the project period.			
6	6	All stages	**CO-ORDINATION OF DUTY HOLDERS**	The client (and all other duty holders) must co-ordinate their activities in respect of those carrying out or affected by construction work. The concept of the lead designer and main contractor is useful in this respect.	Particularly at the start of the project but actively throughout the project period.			
7		All stages	**IDENTIFICATION**	There is a continuing need to identify all those with whom the client needs to co-ordinate activities. This could well relate to overlapping projects, etc.	Particularly at the start of the project but actively throughout the project period. Another project dynamic that must be continuously fulfilled.			
8	7	All stages	**GENERAL PRINCIPLES OF PREVENTION**	The basic tenets of health and safety management must be embraced by all duty holders. Emphasis is duly placed on those designing, planning and involved in the preparation of a project. This regulation has more relevance for those clients undertaking a 'hands-on' approach. Compliance with this regulation will normally be delivered by the competent team assembled by the client.	Throughout the project period.			
9	9	All stages	**MANAGEMENT ARRANGEMENTS**	The client must ensure that management arrangements are effectively set up by duty holders and remain effective throughout the project. Thus roles and responsibilities as well as communication networks must be clear and well understood for the project duration. The client has a duty to monitor and review such arrangements through an amalgamation of checking and seeking confirmation procedures. This carries an over-arching duty, and co-operation and co-ordination duties must be covered in agenda items that need to focus on relevant issues.	Particularly at the start of the project but actively throughout the project period. The client needs assurance before the construction phase starts that co-operation/co-ordination arrangements are suitable, even on a one-day project.			

Continued

Contract

No.	Reg.	Stage	Procedure	Description	Timing of action	Action req. Yes	Action req. No	Initials
				The client must be seen to have asked the right question of the right people at the right time. The client has to be satisfied that arrangements made by the competent duty holder continue to reflect the ongoing competent approach or otherwise take the necessary action. Particular emphasis is given to: • adequate protection for workers and the public • adequacy of welfare arrangements • suitable co-operation/co-ordination between designers and contractors • ongoing effectiveness of contractors' safe and suitable systems of work. **THIS REFLECTS THE CLIENT'S INFLUENCE OVER THE ENTIRE PROCESS**	This is a project dynamic and roles/responsibilities/ communication links will all need to constantly respond to the changing nature of the duty holder interfaces. The client has an over-arching responsibility to ensure that the management arrangements of all duty holders remain effective throughout the project. Progress meeting agendas can provide the means of re-assurance that systems are effective. Audits and inspection summaries need to feed into such meetings. More complex projects need more sophistication in ensuring that effective management arrangements remain in place.			
10	10	All stages	PRE-CONSTRUCTION INFORMATION	Such information must be given to every designer and those contractors appointed or who may be appointed by the client. The client has no such obligation to contractors appointed by third parties. Such relevant information includes: • that about/affecting the site/construction work • proposed use of the structure as a workplace • mobilisation period for contractors • existing health and safety file information. **Note:** This is pre-construction information and could be required during the construction phase (Refer to Section 9 Documentation, pages 137 to 147)	As soon as possible but before related work is undertaken. Such information could be provided in the construction phase and relates to information required before the construction activity itself is undertaken.			
11	12	Pre-initial design	DESIGNS OUTSIDE GREAT BRITAIN	If design is undertaken outside Great Britain the client could be responsible for ensuring that such design is compatible with the requirements of Regulation 11. This would depend on the commissioning chain.	This should be covered by an early question to the design team.			
12	25(2)	Construction phase	CONTROL OF CONSTRUCTION WORK	Any client influencing the way in which a contractor works could incur health and safety management duties relating to site activities (Regulations 25 to 44). Thus the imposition of methodologies on the contractor should be avoided wherever possible and unless absolutely necessary for operational reasons.	Construction period.			
13		Post-construction period.	HEALTH AND SAFETY FILE MANAGEMENT	Even on smaller projects construction work undertaken of a non-notifiable nature could still lead to an amendment/modification to an existing health and safety file. The client's management systems must be capable of responding.	Construction period.			

CLIENT CHECKLIST (Sheet 2 of 2)

4.4
CLIENT CHECKLIST: NOTIFIABLE PROJECTS

CLIENT CHECKLIST: NOTIFIABLE PROJECTS (Sheet 1 of 3)

Additional requirements for all construction projects of notifiable status.

Contract

No.	Reg.	Stage	Procedure	Description	Timing of action	Action req. Yes	Action req. No	Initials	Date actioned
A	14(1)	Post initial design/ pre significant detailed design	APPOINT CDM CO-ORDINATOR	The client must appoint a CDM co-ordinator for every notifiable project, subject to competence and resource adequacy having been established. The client can rely on the advice and assistance of the CDM co-ordinator in fulfilling his duties under the regulations.	As soon as practicable after initial design. If there is any doubt the appointment should be made sooner rather than later.				
B	14(2)	After the appointment of the CDM co-ordinator	APPOINT PRINCIPAL CONTRACTOR	The client must appoint a principal contractor for every notifiable project, subject to competence and resource adequacy having been established. The need to appoint should have been affirmed by the design teams. Both the above appointments must be in writing.	The actual timing of this appointment will reflect the procurement strategy, but there is strong emphasis that the appointment should be made early enough to provide the necessary perspective for constructability and planning/ managing the project.				
C	4(1)(a) and 4(1)(c)		ESTABLISH COMPETENCE	The client must be satisfied that the person/ organisation appointed has the necessary competence to perform the functions of CDM co-ordinator and principal contractor, and has the organisational structure to ensure that individuals are either competent or under the supervision of a competent person.	Prior to appointment.				
D	9(1)	Post initial design/ pre significant detailed design	ENSURE ALLOCATION OF SUFFICIENT TIME AND RESOURCES	Similarly there is the need to ensure that adequate time and resources are allocated to perform the functions of CDM co-ordinator and principal contractor.	Prior to appointment.				
E	14(4)(a) and 14(4)(b)	Post initial design/ pre significant detailed design	CONFIRM EXTENT OF CDM CO-ORDINATOR SERVICE TO BE PROVIDED	All CDM co-ordinator duties must be fulfilled either by the client or by a third party (internal or external) appointed by the client. No CDM co-ordinator or principal contractor work should be undertaken until the appointments have been verified in writing.	Prior to appointment. If appointments are not made at the appropriate time, or the service to be undertaken is not fully defined, the client will be deemed to function as the CDM co-ordinator and principal contractor.				
F	14(5)	Post initial design/ pre significant detailed design	APPOINTMENTS OF CDM CO-ORDINATOR AND PRINCIPAL CONTRACTOR TO BE MADE IN WRITING	This applies to in-house and external appointments alike.	Simultaneously with appointment.				

Continued

Continued

Contract						Action req.			
No.	Reg.	Stage	Procedure	Description	Timing of action	Yes	No	Initials	Date actioned
G	*Schedule 1*	Post initial design/ pre significant detailed design	**SIGN THE RELEVANT PART OF THE NOTIFICATION TO THE HSE OR THE OFFICE OF RAIL REGULATION (F10)**	The client or someone on his behalf must sign to confirm that the client understands his duties under the CDM Regulations 2007. This must be someone strategically placed within the client's organisation who can sign on the client's behalf. It should not be signed purely for convenience.	As soon as practicable after the appointment of the CDM co-ordinator. This document should be provided to the client by the CDM co-ordinator, whose further role is to ensure it is sent off to either the HSE or the Office of Rail Regulation.				
H	15	On appointment	**PROVIDE THE CDM CO-ORDINATOR WITH PRE-CONSTRUCTION INFORMATION**	This information has to be identified and collected by the CDM co-ordinator. It is information which is in the client's possession or reasonably obtained by the client. Such information includes the stipulated minimum amount of time for planning and managing before construction work starts. Thus on notifiable projects the CDM co-ordinator will facilitate a number of the duties to be discharged by the client on all projects. However, this does not translate into being a transference of legal duty.	Such information must be provided promptly.				
I	16	Pre-construction phase	**CONTROL THE START OF CONSTRUCTION**	This is a primary control gateway to be exercised by the client, who must be satisfied that the principal contractor has developed the construction phase plan compatible with the basic obligation to ensure that work activities can be undertaken without risk to health and safety to those undertaking construction and those who might be affected by it. The CDM co-ordinator must be available to assist the client in this respect. This ensures that the principal contractor is adequately prepared and that the start of construction is not premature due to the perceived need for an early start without due preparation of controls.	Prior to the start of the construction phase. This includes enabling works and any works for the preparation of the structure, including site clearance, etc. (refer to Interpretation, Regulation 2) **Note:** Failure to ensure compliance with Regulation 16 confers a right of civil action.				

CLIENT CHECKLIST: NOTIFIABLE PROJECTS (Sheet 2 of 3)

Contract					Action req.		Initials	Date actioned
No.	Reg.	Stage	Procedure	Description	Timing of action	Yes	No		

No.	Reg.	Stage	Procedure	Description	Timing of action	Yes	No	Initials	Date actioned
J	17	From the appointment of the CDM co-ordinator onwards	**HEALTH AND SAFETY FILE MANAGEMENT**	The client must provide the CDM co-ordinator with relevant information in his possession or reasonably obtained by him needed for inclusion in the health and safety file. Particular mention is made of the Control of Asbestos Regulations 2006.	Immediately such information is available.				
				The CDM co-ordinator identifies such information for the client and subsequently collects it.	Soon after the appointment of the CDM co-ordinator.				
K	17(2)		HEALTH AND SAFETY FILE FORMAT	The CDM co-ordinator should discuss at an early stage the form and format of the health and safety file with the client. This becomes a client document after eventual handover and hence the client should be involved in its compatibility with existing management systems, computerised or otherwise.	Soon after the appointment of the CDM co-ordinator.				
					At the end of the construction phase. Essentially when the insurance arrangements for the project transfer from the principal contractor's cover to that of the client.				
				Information in a single file must be easily identifiable.					
				The client should also expect to receive a proportional health and safety file in respect of partial handover and sectional completion, since insurance cover has also been transferred.					
L			RECEIPT OF HEALTH AND SAFETY FILE	The CDM co-ordinator must hand the health and safety file to the client and should receive a signature for its delivery and acceptance.	An explanatory letter should accompany the health and safety file.				
M	17(3)	Post-construction operation, maintenance and demolition/ dismantling.	**MAINTAIN AVAILABILITY AND CURRENCY OF THE HEALTH AND SAFETY FILE**	The information within the health and safety file must be kept up to date and available for inspection by any person who might reasonably need that information.	This is an ongoing duty and relates to maintaining the security of access to the health and safety file as well as managing future amendments/ modifications to any existing health and safety file.				
N	17(4)	Change of ownership	**TRANSFER OF OWNERSHIP**	The health and safety file must accompany change of ownership together with an explanation of the nature and purpose of the health and safety file for the benefit of the new owner.	Simultaneously with conveyancing procedure.				

Section 5
THE DESIGNER

5.1 Introduction

The design team occupies a unique position in the construction process not only as professional advisor to the client but also through their strategic position near the top of the supply chain. The Health and Safety Executive (HSE) acknowledge this and continue to emphasise the influential position of design in the delivery of effective health and safety management.

The CDM Regulations 2007 require much less of a cultural step from the designer than that imposed on the client in the fulfilment of duties. Co-operation (Regulation 5) and co-ordination (Regulation 6) are there to be discharged, as are the general principles of prevention (Regulation 7). However, these regulations reinforce the good practice approach of old and the standard tenet of health and safety management generally discharged by the practitioner for some time.

The main areas of change surround:

- the competence of the designer
- the process of design demonstration
- designing for the provisions of the Workplace (Health, Safety and Welfare) Regulations 1992
- the interface between designer, CDM co-ordinator and principal contractor in respect of design change.

The definition of design now includes:

- drawings
- design details
- specification and bill of quantities (including specification of articles or substances) relating to a structure
- calculations prepared for the purpose of design
- the need in designing a structure as a place of work to account for the provisions of the Workplace (Health, Safety and Welfare) Regulations 1992.

Designer itself means:

'any person (including a client, contractor or other person referred to in these Regulations) who in the course or furtherance of a business:

a)   prepares or modifies a design; or

b)   arranges for or instructs any person under his control to do so relating to a structure or to a product or mechanical or electrical system intended for a particular structure, and a person is deemed to prepare a design where a design is prepared by a person under his control.'

The designer's perspective embraces all those foreseeable aspects of health and safety associated with constructability, useability, maintainability and replacement up to and including demolition and dismantling. These duties to be discharged by the designer apply to both permanent and temporary design.

Regardless of their pedigree, designers must consider the need to design in a way which avoids risks to health and safety or reduces those risks as far as is reasonably practicable so that the projects they design can be constructed, used, operated, maintained, replaced and taken down safely. Where risks cannot be avoided, suitable and sufficient information on them has to be provided to those who need the information. Designers must therefore be able to:

- identify the hazards inherent in their designs
- identify the resultant risks from construction through to demolition
- understand how to eliminate the hazards or reduce the risks
- communicate relevant information to others.

Fundamentally, designers must not produce designs that cannot be constructed, used, maintained, replaced and removed safely.

Much of design is interdisciplinary and the process itself requires co-ordination. This is addressed in Regulation 6, with the ACoP promoting the concept of the lead designer. Whilst there is no legality to this position it is a legal requirement that duty holder roles must be co-ordinated. Hence, the lead designer as the point of co-ordination should be identified on all projects.

Designers are more comfortable with the interpretation of their role after the 12-year experience offered under the CDM Regulations 1994. The HSE's interventionist strategies associated with raising the awareness of the designer in the discharge of duties were helpful but have now moved away from the demands of form-filling evidence to sensible realism and focus on managing risk and not paperwork.

Designers own their process and should therefore be capable of demonstrating their contribution to effective health and safety management through the visibility of their systems rather than demonstrability with detail at every stage. Indeed the most effective design contributions come from strategies developed with other duty holders and in particular the integrated contribution between design and construction. Within the context of project management it is the team that delivers success rather than one party working in isolation. The same is true within the health and safety environment and designer–contractor–end user interfaces and perspectives remain critical.

Paragraph 144 of the ACoP reminds us that there is no legality in the form of the records kept, although it is useful to have some record. This needs to be balanced against the information overload that accompanies excess paperwork or bureaucratic responses and the danger of losing key messages because of the irrelevance of much that accompanies that excess. The communication of relevant information is based on the concept of the competence of other duty holders, particularly the contractor, and the need to communicate in areas of significance rather than the mundane. Accompanying this approach must also be the awareness that inadvertently this is a political arena and the design process must be seen to have addressed those issues that are the concern of the HSE. This concern is based on statistical feedback and therefore it is sensible to give visibility in the design process to those valid areas of address without succumbing to the gross excesses of the past, which serve no one's best interests.

Associated perspectives must embrace not only safety but also related occupational ill-health issues, including:

- musculoskeletal disorders
- noise
- respiratory illness
- sensitory dermatitis
- hand/arm vibration syndrome.

This is not an exhaustive list and it will need to be extended to capture the specifics of any project.

For many years the design fraternity has made wide use of the design risk assessment pro forma and all its variations. Whilst these can be useful, design itself has recourse to other means of demonstrating its process from numerous systems/techniques/approaches, including:

- design review meetings
- brainstorming sessions
- team meetings
- workshops
- notes on drawings
- project risk (health and safety) registers
- accompanying minutes.

Indeed, there is much merit in using existing project management techniques where they can accommodate the shared objectives of CDM management. In that sense the project risk (health and safety) register offers a facility that more closely represents the iterative nature of the design process and provides a broad canvas to which design teams and others contribute.

The example project given in Chapter 9 illustrates the main characteristics of such a register, namely:

- integrated document with contributions from all duty holders
- the promotion of a team-based approach
- a dynamic approach and issues kept alive until delivered
- a move away from pseudo-numerical classifications towards general categories
- visibility throughout to significant issues
- documentation accompanies the project from start to finish.

Such registers are in common use by project management teams (mainly for aspects of financial exposure and control) and if health and safety is to be truly integrated into the considerations of the team then an extension of standard procedures to accommodate health and safety demonstrability and communication has much merit. The remaining question, however, relates to the ownership of the register. This should not be difficult – it could be owned by the project manager, lead designer and/or the CDM co-ordinator as appropriate.

Figure 5.1

REGULATIONS TO
BE DISCHARGED
BY DESIGNERS

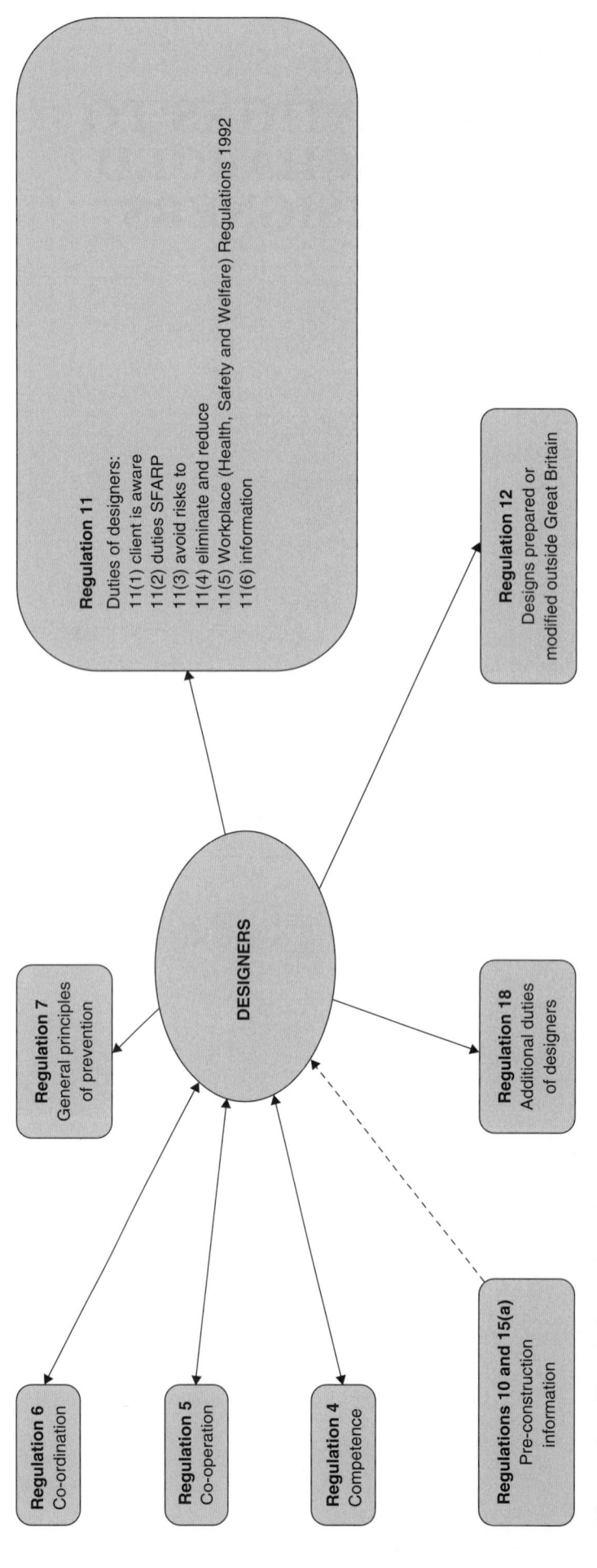

Figure 5.1 Regulations to be discharged by designers

5.2 All projects

For all projects designers must:

- ensure they are competent and adequately resourced
- co-operate with other duty holders on this and adjoining sites
- co-ordinate their activities to improve the management /control of risk
- design in conjunction with the general principles of prevention
- ensure client is aware of his duties
- discharge a hierarchal approach to health and safety contribution
- account for the provisions of the W(H,S&W) Regulations 1992
- provide sufficient (and suitable) information about aspects of design to:
  - clients
  - other designers
  - and contractors.

As such the process must achieve integration between the various elements of design including clients scope of works and other relevant stakeholder interest groups, permanent and temporary design, work specialist package design and sub-design generally. As with all other duty holders, focus cannot solely be on the specifics of the CDM Regulations 2007, for it is the whole portfolio of construction related legislation that is relevant to the task that needs to be embraced.

Throughout the Regulations duty holders have to discharge duties:

'so far as is reasonably practicable.'

The Court of Appeal's interpretation of this phrase is:

'Reasonably practicable is a narrower term than physically possible, and implies that a computation must be made in which the quantum of risk is placed in one scale, and the sacrifice, whether in money, time or trouble, involved in the measures necessary to avert the risk, is placed in the other; and that, if it be shown that there is a gross disproportion between them, the risk being insignificant in relation to the sacrifice, the person upon whom the duty is laid discharges the burden of proving that compliance was not reasonably practicable. This computation falls to be made at a point of time anterior to the happening of the incident complained of.'

On every project the design team must be able to demonstrably answer the question:

'what is your contribution to health and safety management on this project?'

How this strategy is implemented is judgmental but like other related processes it must be demonstrable. The focus at all times is towards significant (and principal) issues specific to the site, based on the concept of the competent contractor.

For the designer, occupying such an influential position in the health and safety management chain, an appropriate hazard identification and management strategy is fundamental. All stages of design starting with initial design must embrace this holistic concept and deliver an appropriate strategy (See the holistic diagram on page 22).

Principally, the purpose of a hazard identification and management strategy in relation to health and safety follows a simple audit trail of:

- Identification
- Evaluation
- Contribution
  1. Elimination
  2. Minimisation
  3. Transfer
- Monitor
- Review.

5.3   Identification

This is the first step in the health and safety risk management process and if team members are unable to identify the issues either through ignorance or complacency the succeeding elements are flawed. Often it involves a lateral perspective allowing freedom to think outside the 'box' and hence the process is not facilitated from a checklist approach. The competent team should have no problem in identifying the salient health and safety management issues.

5.4   Evaluation

All evaluation processes are challenged in respect of their validity in the process of design contribution. The method of hazard evaluation or appraisal after identification is not prescribed and ranges from the simple qualitative description of high, medium or low to the quasi-numerical approach based on a 10×10 matrix accounting for likelihood and severity. More sophisticated methods can be used such as HAZOP (Hazard and Operability Studies) and probabilistic methodologies but these are not often required for main stream design. Most evaluation methods provide a ranking for the prioritisation of action by the design team. What is critical is that the hazard is identified and a design contribution made based on a hierarchy of prevention and protection. There remains a strong argument that the evaluation stage can be omitted with no detriment to the overall process for the designer has already exercised judgement in listing the issue.

5.5   Contribution

The design team is aware of the need to identify the issues and having identified them are then challenged to make a health and safety contribution based on the following hierarchy of response in respect of the content of regulation 7 and its linkage with Appendix 7 (of ACoP):

- Elimination
- Minimisation/reduction
- Transfer

It is vital not to embark on a generic approach (anathema to the Health and Safety Executive[HSE]) for today's design team challenge is to think 'outside the box'. The checklist mentality only ensures that the designer becomes a prisoner of the checklist boundaries, which stifles creativity and health and safety management contribution.

Design teams must be aware of the health and safety management issues before they are in a position to proactively contribute. That awareness must extend to a competent appreciation of relevant health and safety issues aligned to the focus projected by any current HSE interventionist strategy. This must also be a perspective from which CDM co-ordination approaches the interface with design.

Thus, design teams must:

- consider
- contribute
- communicate
- co-ordinate.

in a health and safety management process, which engages all design team members, and includes design changes arising from variation orders and architects instructions.

The transfer of residual information down the supply chain is part of the communication link and the ACoP informs that this can be achieved by:

- notes on drawings
- written information provided with the design
- suggested construction sequences.

A visible system based on appropriateness and minimisation of paperwork must accompany the design process to ensure transparency and accountability and to provide documentary evidence that duties have been fulfilled without perpetuating the bureaucratic nightmare associated with self-generating paperwork systems.

ACoP paragraph 145 reminds us all that

'Too much paperwork is as bad as too little, because the useless hides the necessary. Large volumes of paperwork listing generic hazards and risks, most of which, are well known to contractors and others who use the design are positively harmful, and suggest a lack of competence on the part of the designer.'

Documentary evidence should merge with existing management systems and could be demonstrated in a variety of forms, including:

- agenda items
- minutes of meetings
- design office manuals
- project risk registers
- design philosophy statements.

However in establishing a trail of accountability, management control systems must support the practitioner in discharging legal duties without enslaving him/her to a bureaucratic response.

Further to the contribution made, there is a design team requirement to ensure that adequate information is provided to others so that work at all stages (constructability; operability; facilities management; and de-commissioning) can be undertaken safely. Thus appropriate outputs must be delivered into the pre-construction information and the health and safety file process at the appropriate time.

This should also focus attention on design liability associated with the extended duration of health and safety files as well as amendment/modifications requirements to existing files on the non-notifiable project.

The designer must be aware of the need to provide information to clients, other designers and contractors as part of the information flow process. This process is iterative and information must flow out of design to ensure that work can be carried out safely at the appropriate time

5.6 Design change

Design changes during the construction phase are usually initiated by variation orders (VOs) or architect's instructions (AIs). Where such instructions are received during the construction phase they can have greater implications within the confines of a construction programme.

These instructions will also need to have been subjected to the ongoing design risk management strategies and supporting documentation will provide the information needed.

Figure 5.2
SYSTEMS APPROACH FOR THE DESIGNER

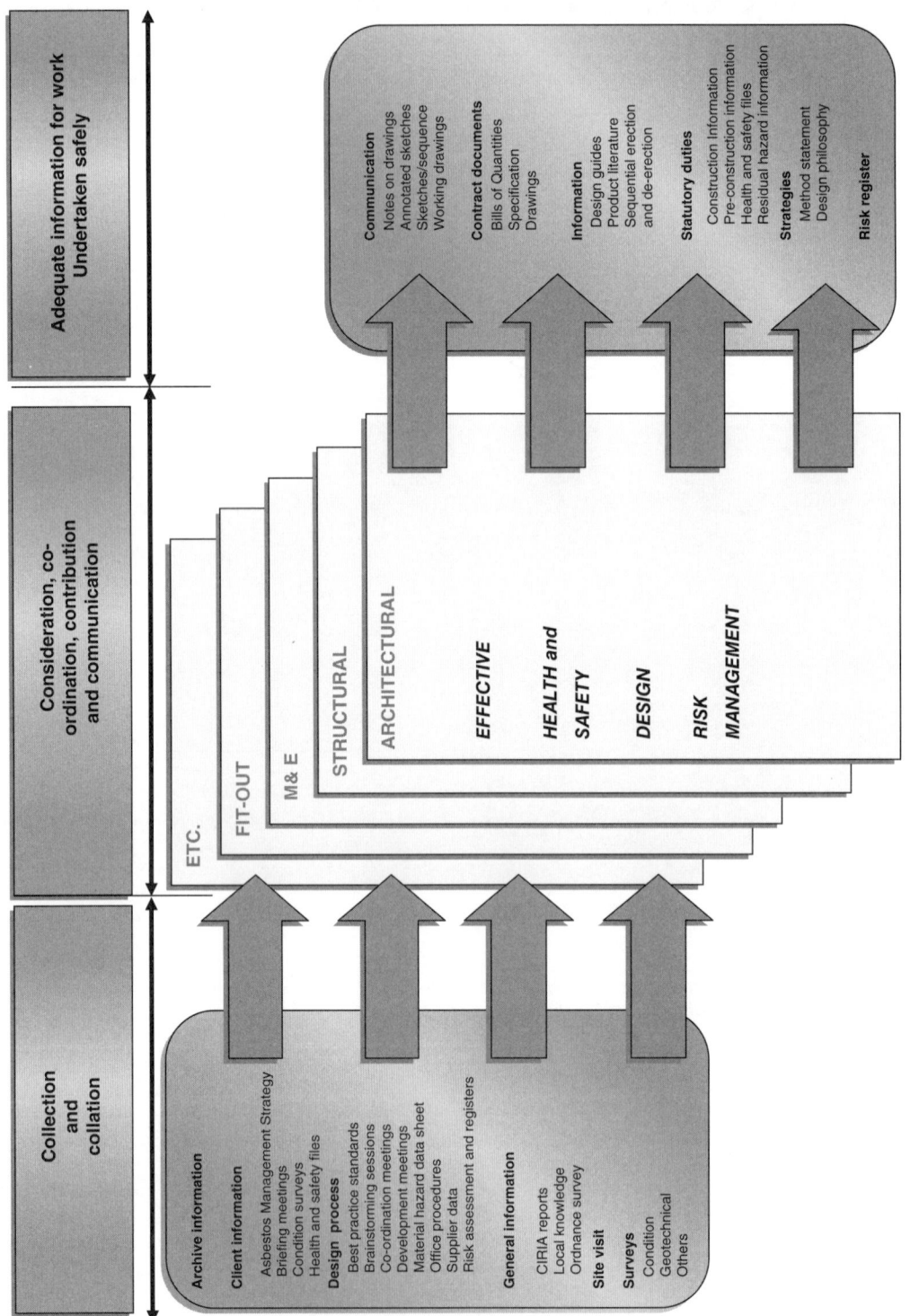

Figure 5.2 Systems approach for the designer

5.1
DESIGNER FLOWCHART: ALL PROJECTS

Initial design phase	Significant detailed design phas	Pre-construction phase	Construction phase	Post-construction phase
Appoint competent duty holders *4(1)*	**Appoint competent duty holders** *4(1)*	All duty holders whenever appointed must be assessed for competence. In accepting the design team self-certify their competence.	No person appointing can rely on self-certification arising out of regulation 4(1)(b).	**Note: Depending on the procurement strategy, design duties can extend into construction phases.**
Client to be made aware of his duties *11(1)*	Imperative that before design starts client is made aware of client's duties. There are no exceptions.	Where numerous designers are involved it is sufficient if they ensure that the lead designer has informed the client of his duties.		
Designs prepared or modified outside Great Britain *12*	Designers who commission designs to be prepared or modified outside Great Britain must ensure that all designers discharge duties compliant with regulation 11.			
Receipt of pre-construction information *10(1)(a) or 15(a)*	Information is provided to all designers either directly from the client (non-notifiable project) or from the CDM co-ordinator (notifiable project).	Such information is essential for the design focus on related issues and will be distributed compatible with design appointments.	As such it is not limited solely to phases prior to the start of construction, since certain design packages will only be appointed during the construction phase itself.	Such information includes: ● information about the site ● proposed use of the structure ● mobilisation period ● existing health and safety file information.
Co-operate and seek co-operation *5(1)*	**Co-operate and seek co-operation** *5(1)*	**Co-operate and seek co-operation** *5(1)*	**Co-operate and seek co-operation** *5(1)*	All duty holders must co-operate with and actively seek the co-operation of those involved in construction on the same or adjoining sites. **NOTE:** The project site is also an adjoining site to those around. This will remain a dynamic situation throughout the project.
Co-ordination *6*	**Co-ordination** *6*	**Co-ordination** *6*	**Co-ordination** *6*	The design process must be duly co-ordinated and the ACoP recommends the concept of the lead designer to facilitate and own this process which is another project dynamic.
General principles of prevention *7*	This is the guiding principle for the discharge of duties under regulations 11(2); 11(3) and 11(4).	The basic tenet of health and safety management. As outlined in the MHSW Regulations 1999.		

Initial design phase	Significant detailed design phas	Pre-construction phase	Construction phase	Post-construction phase
As far as is reasonably practicable *11(2)*	Essential that health and safety does not dominate the design process but is taken as an integral and balanced consideration.	Quantum of risk approach in the discharge of duties compatible with the term '*as far as is reasonably practicable*'.		
Avoid foreseeable risks *11(3) & 11(4)*	**Avoid foreseeable risks** *11(3) & (4)*	Such risks must account for those carrying out construction work; liable to be affected by it; cleaning; maintaining and using the structure as a workplace.	General principles of prevention must be embraced in the discharge of these duties (Appendix7). Thus collective measures have precedence over individual measures.	
Use as a place of work. *11(5)*	**Use as a place of work.** *11(5)*	Workplace (Health, Safety and Welfare)Regulations 1992 must be complied with for all fixed workplaces. This applies to the design of offices, shops, factories and schools. Many designs would not have to subscribe to these regulations.	Account must be taken of the provisions of these regulations in relation to the design of and materials used in the structure.	This has impact on design inputs to both pre-construction information and the health and safety file.
	Sufficient information *11(6)*	Adequate information for the corresponding work of to be undertaken safely must be provided to clients; other designers and contractors. Such information relates to relevant aspects of design of the structure, its construction and/or maintenance.	This information is required as inputs to associated documentation such as pre-construction information; drawings; health and safety files etc.	
			Designers need to be aware that as a result of the non-notifiable project amendments/ modifications might be required to an existing health and safety file.	Modifications to an existing health and safety file

5.2
DESIGNER FLOWCHART: NOTIFIABLE PROJECTS, ADDITIONAL DUTIES

Initial design phase	Significant detailed design phase	Pre-construction phase	Construction phase	Post-construction phase
Appointment of CMD co-ordinator *18(1)*	CDM co-ordinator must be appointed as soon as practicable after initial design.* All designers must ensure that this has been done. Failure to appoint as soon as practicable after initial design ensures that the client takes on the duties of the CDM co-ordinator.	This duty must also link with the requirement to ensure the client is aware of his duties under Regulation 11(1). The client's understanding of such duties can be gauged by client performance. If this is not satisfactory then appropriate action must be taken.	Appropriate action would require further advice and information to be given to the client.	
Provide sufficient information with design *18(2)*		This information is to be provided to the CDM co-ordinator and relates to relevant aspects of the design of the structure or its construction or maintenance to enable the CDM co-ordinator to distribute Pre-Construction Information and prepare the health and safety file	**Note**: The design duties to comply with the Workplace (Health, Safety and Welfare) Regulations 1992 will impact on design information required for the health and safety file.	
			Health and safety file *18(2)*	Ensure information such as 'as built' drawings etc and residual hazard information is provided to the CDM co-ordinator to expedite the handover of the health and safety file at construction completion.

*Paragraph 66 of the ACoP distinguishes between initial design and significant detailed design. Significant detailed design includes: preparation of the initial concept design and implementation of any strategic brief. Hence the design process does not have to proceed far before the appointment of the CDM co-ordinator needs to take place, i.e. sooner rather than later.

5.3
DESIGNER CHECKLIST: ALL PROJECTS

DESIGNER CHECKLIST: ALL PROJECTS (Sheet 1 of 4)

Contract

No.	Reg.	Stage	Procedure	Description	Timing of action	Action req. Yes	Action req. No	Initials
1	*4(1)(a)*	Pre-appointment	**APPOINT COMPETENT DUTY HOLDERS**	All duty holders, i.e. designer(s) and contractor(s) appointed directly by the designers, must be assessed as competent. The designer must be satisfied that the person/ organisation appointed has the necessary competence to perform the function. The designer cannot rely on any party self-certifying themselves as competent. The rigour of the assessment must relate to the complexity of the project.	Such assessment must be undertaken before appointment.			
2	*4(1)(b)*	Pre-appointment	**COMPETENCE**	In accepting the design appointment do you have the competence to fulfil the role on this project?	In accepting the party self-certifies he is competent			
3		Pre-appointment	**IDENTIFY DUTY HOLDERS**	To fulfil the above duty the designer must identify all designers and contractors who will be appointed by him. This reflects the procurement strategy of the project.	Throughout the project.			
4	*11(1)*	Post-design appointment	**ENSURE CLIENT IS AWARE OF HIS DUTIES**	All designers have a duty to ensure that the client is aware of his duties. There are no exceptions. On larger projects it would be sufficient if each designer received confirmation that this had been undertaken by the lead designer.	Immediately on the appointment of the design team.			
5	*12*	Design stage	**DESIGNS PREPARED OR MODIFIED OUTSIDE GREAT BRITAIN**	Designers who commission designs to be prepared outside Great Britain must ensure that all designers discharge duties compliant with Regulation 11.	Before commissioning.			
6	*10(1)(a) or 10(1)(b)*	Design stage	**RECEIVE PRE-CONSTRUCTION INFORMATION**	This should be forwarded either by the client (non-notifiable project) or the CDM co-ordinator (notifiable project).	Before design starts.			
		Before the relevant stage of design	**RECEIPT OF PRE-CONSTRUCTION INFORMATION**	Pre-construction information is to be issued promptly to all those designing the structure.	To enable the designer to effectively embrace relevant information within his design process.			

Continued

Contract						Action req.		
No.	Reg.	Stage	Procedure	Description	Timing of action	Yes	No	Initials
7	5	All stages	**CO-OPERATION OF DUTY HOLDERS**	The designer and all other duty holders must seek the co-operation of those involved in construction not only on the project but also on adjoining sites. There is a further need to co-operate with those involved in construction not only on the project but also on adjoining sites. Co-operation is a fundamental of effective project management. **Note:** Your project is itself an adjoining site.	Particularly at the start of the project but actively throughout the project period. The more complex the project the greater the co-operative effort required. The continuing need for co-operation ensures that relevant parties move into and out of the project throughout its duration.			
8		All stages	IDENTIFICATION	There is a continuing need to identify all those with whom the designer needs to co-operate. This could well relate to overlapping projects, etc.	Particularly at the start of the design process but actively throughout the project period.			
9	6	All stages	**CO-ORDINATION OF DUTY HOLDERS**	The designer (and all other duty holders) must co-ordinate their activities in respect of those carrying out or affected by construction work. This is facilitated by the concept of the lead designer.	Particularly at the start of the design process but actively throughout the project period.			
10		All stages	IDENTIFICATION	There is a continuing need to identify all those with whom the designer needs to co-ordinate activities. This could well relate to overlapping projects, etc. Emphasis is placed on the role of the lead designer in the process of co-ordination. Co-ordination, like co-operation and communication, is an essential component of effective project management.	Particularly at the start of the design process but actively throughout the project.			
11	7	All stages	**GENERAL PRINCIPLES OF PREVENTION**	The basic tenets of health and safety management must be embraced by all duty holders. Such compliance also links with design duties particularly under Regulations 11(2), 11(3), 11(4) and 11(6).	Throughout the project period.			
12	11(2)	Design	**AS FAR AS IS REASONABLY PRACTICABLE**	The designer is reminded that the design process is not meant to be dominated by health and safety considerations. Health and safety should remain an integral consideration along with other design issues such as form, function, aesthetics, environmental impact and other relevant construction-related legislation.	Throughout the design process.			

66

Contract

No.	Reg.	Stage	Procedure	Description	Timing of action	Action req. Yes No	Initials
13	11(3)	Design	**AVOID FORESEEABLE RISKS TO HEALTH AND SAFETY AFFECTED BY THE CONSTRUCTION PROCESS**	Foreseeable risks must be avoided to those: • carrying out construction work • liable to be affected by such construction work • cleaning windows, transparent/translucent walls, ceilings or roofs • maintaining fixtures/fittings • using the structure as a workplace. This must be undertaken to discharge duties as far as is reasonably practicable based on a quantum of risk approach, i.e. a disproportionate response is not required to remove minor issues.	Throughout the design process.		
14	11(4)	Design	**HIERARCHICAL APPROACH BASED ON:** • **ELIMINATING HAZARDS** • **REDUCING RISKS FROM REMAINING HAZARDS** **ALL WITH RESPECT TO THE TERM 'AS FAR AS IS REASONABLY PRACTICABLE'** This requires an embrace of the basic tenets of health and safety management as outlined in Appendix 7.	This is achieved by the design process fulfilling a hazard identification and management strategy based on the principles of prevention and protection. The design process should focus on managing risk and not the undue management of paperwork. Whilst documentary evidence of the process is required the designer should note that the strategy is evidenced through a portfolio of design aspects, including: • the design office manual • notes on drawings • project risk registers • design risk assessments. Such an approach must be taken with all aspects of design, including variation orders and architects' instructions. It must to accepted that late design change has a great impact on the controls already set up by the contractor and the full effects of change must be considered.	Throughout the design process.		

Continued

Contract						Action req.		Initials
No.	Reg.	Stage	Procedure	Description	Timing of action	Yes	No	
15	11(5)	Design	**USE AS A WORKPLACE**	Account must be taken of the Workplace (Health, Safety and Welfare) Regulations 1992. These relate to the design of fixed workplaces such as offices, shops, factories and schools. Much design is done outside these areas and would therefore not have to comply.	Throughout the design process.			
16	11(6)	Design construction	**SUFFICIENT INFORMATION TO ADEQUATELY ASSIST OTHERS IN COMPLYING WITH THEIR DUTIES**	This applies to the provision of adequate information about aspects of design, construction and maintenance to: ● clients ● other designers ● contractors to enable them to discharge their duties under the regulations. Suitable and sufficient information must be communicated for the purpose of effective health and safety management. Designers need to be aware that even for non-notifiable projects management systems must respond and amendments/modifications might be required to an existing health and safety file.	Throughout the design process and therefore would include information necessary for the flow into pre-construction information as well as the necessary inputs to: ● tender documentation ● notes on drawings ● project risk registers ● amendments/modifications to existing health and safety files.			

5.4
DESIGNER CHECKLIST: ADDITIONAL DUTIES FOR NOTIFIABLE PROJECTS

DESIGNER CHECKLIST: ADDITIONAL DUTIES FOR NOTIFIABLE PROJECTS

Contract

No.	Reg.	Stage	Procedure	Description	Timing of action	Action req. Yes	No	Initials	Date actioned
A	18(1)	Before significant detailed design	**ENSURE A CDM CO-ORDINATOR HAS BEEN APPOINTED**	Design should not pass beyond the initial design stage unless a CDM co-ordinator has been appointed. **Note:** Such appointments by the client must be in writing. **Note:** An incorrect interpretation of the demarcation between initial design and significant detailed design will place the client in the role of the CDM co-ordinator.	As soon as practicable after initial design.				
B		Throughout design	**CO-ORDINATION**	Establish effective co-ordination systems between all designers. This should be effected by the lead designer. **Note:** This is a project dynamic with co-ordination required between all designers throughout the project. Particular emphasis needs to be placed on the interface between permanent and temporary design.	Particularly at the start of the design process but throughout design and construction up to and before project handover.				
C		Before significant detailed design	**RECEIVE PRE-CONSTRUCTION INFORMATION**	This should be provided to every designer by the CDM co-ordinator.					
D	18(2)	Throughout design	**PROVIDE SUFFICIENT INFORMATION**	Such information about aspects of the design of the structure or its construction or maintenance as is necessary for the CDM co-ordinator to provide further pre-construction information to other designers and contractors who have been or who might be appointed by the client. It should include relevant information for input into the health and safety file such as: • residual hazards • maintenance methodologies • 'as-built' record drawings.	Throughour design and construction up to and before project handover.				

Section 6
THE CONTRACTOR

Contractors are defined under Regulation 2(1) as:

> 'any person (including a client, principal contractor or other person referred to in these Regulations) who, in the course of furtherance of a business, carries out or manages construction work.'

Contractors are at the exposed end of the construction process and positioned some way down the supply chain. Within this category are specialist contractors and those who themselves manage other contractors (sub-contractors).

Paragraphs 177 and 178 of the ACoP indicate that:

> 'All contractors (including utilities, specialist contractors, contractors nominated by the client and the self-employed) have a part to play in ensuring that the site is a safe and healthy place to work. The key to this is the proper co-ordination of the work, underpinned by good communication and co-operation between all those involved.

> Anyone who directly employs, engages construction workers or controls or manages construction work is a contractor for the purposes of these Regulations. This includes companies that use their own workforce to do construction work on their own premises. The duties on contractors apply whether the workers are employees or self-employed and to agency workers without distinction.'

As with other duty holders all must be competent (Regulation 4), co-operate (Regulation 5), co-ordinate (Regulation 6) and discharge duties compliant with the principles of prevention (Regulation 7). Additional and distinct duties apply beyond this (Regulation 13) and must embrace further specific duties for notifiable projects (Regulation 19).

Contractors operate at the exposed edge of construction and make up the major statistical proportion of site fatalities and associated occupational ill-health and disease. Hence emphasis is placed on the essential nature of management control in respect of safe and suitable systems, and their constant monitoring throughout the construction phase. Irrespective of the sophistication of control mechanisms and accompanying documentation, all exercises remain flawed and fallible if not accompanied by a regular monitoring regime that reviews and refines to ensure effectiveness.

Communication is the essential prerequisite that enables the delivery of successful project management and the contractor must ensure that all employees and relevant others get *the right information to the right people at the right time*. This is usually achieved through information and training delivered via the induction process and this remains one of the vehicles by which some of the duties enshrined in Regulation 13 can be delivered.

It should be noted that the induction programme is not a one-off procedure but must be re-delivered to reflect changes in process and/or procedures.

The contractor, and specifically the sub-contractor, are positioned well down the supply chain and are often abused within the system. As such the sub-contractor is a convenient factor fulfilling the role of financial buffer instead of being employed for the strengths they can undoubtedly bring to the process. They are vulnerable in a process that continually seeks to transfer risk instead of managing it and end up with no option but to attempt to manage the issues handed down by others who are adverse to risk exposures that they are better positioned to deal with.

Whilst the ownership of health and safety should be dealt with by those best positioned to deal with it, a disproportionate response can often be demanded of the contractor. The larger contractor has the power to resist, but the smaller contractor does not.

As noted in Research Report 218:

'almost half of all accidents in construction could have been prevented by designer intervention and that at least 1 in 6 of all accidents are at least partially the responsibility of the lead designer in that opportunities to prevent accidents were not taken.'

The Report identifies that:

'In many instances contractor design incompetence was a major contributor to an accident.'

This latter statement has further resonance in consideration of the popularity of the design and build procurement route.

In addressing some of the above points, duty holder roles have been extended to improve the basics of competence, co-operation and co-ordination, with the client ensuring that management arrangements are effective before work starts and remain so throughout via regimes of monitoring and review.

The boundary conditions of project cost and limited windows of opportunity remain challenges for all and particularly the contractor, but must not be used as mitigation for failure to deploy and maintain effective controls in the harsh environment that is the construction phase.

The cultural movement seen in the last 12 years owes much to the large to medium-sized contractor who has initiated responsible approaches to health and safety, but more could have been done if the construction industry operated within a truly integrated environment.

The aspirations of the Egan Report enshrined in its five key drivers of change, namely *'committed leadership, a focus on the customer, integrated processes and teams, a quality driven agenda and commitment to people'* unfortunately remain as vivid a picture of the future, instead of the present, as they did nearly ten years ago.

Figure 6.1
CONTRACTOR DUTIES: ALL PROJECTS

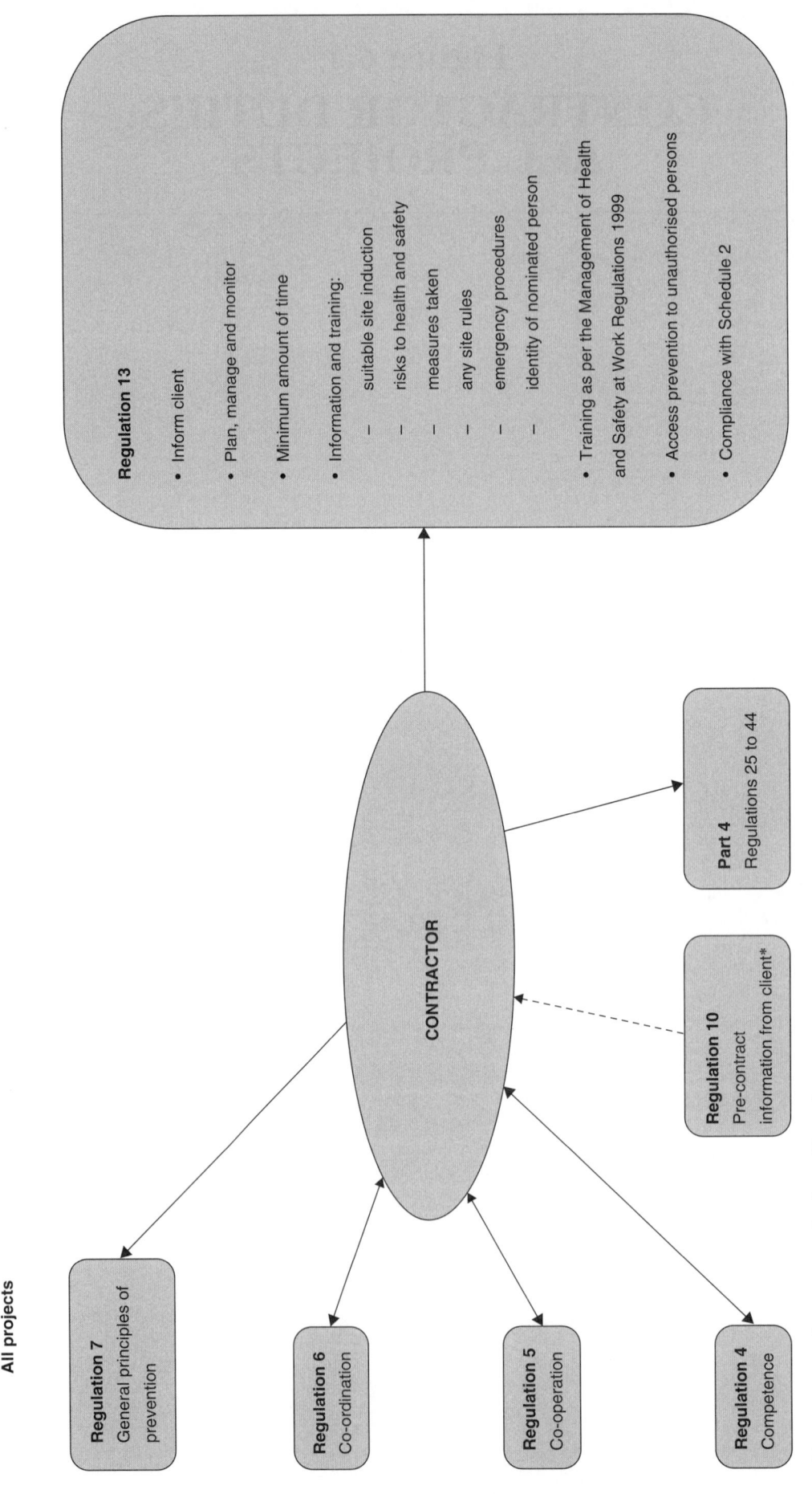

Figure 6.1 Contractor duties: all projects
*To contractors who have been, or might be, appointed by the client.

6.1
CONTRACTOR FLOWCHART: ALL PROJECTS

Initial design phase	Significant detailed design phase	Pre-construction phase	Construction phase	Post-construction phase
		Appoint competent duty holders *4(1)*	**Appoint competent duty holders** *4(1)*	All duty holders whenever appointed must be assessed for competence. In accepting such an appointment the duty holder self-certifies their competence. No person appointing can rely on self-certification arising out of Regulation 4(1)(b).
Note: All contractors who have been or who may be appointed by the client must be in receipt of relevant pre-construction information. Furthermore, any contractors who take on design duties need to familiarise themselves with the section on the designer. This is true of: • the principal contractor managing the design process • specialist work packages involving design elements • anyone who alters a design • anyone who alters a specification.		**Contractor to ensure client is aware of his duties** *13(1)*	No construction work can start unless a client is made aware of his duties. This is a duty that applies to all contractors, regardless of who appoints them.	In practice it would be satisfactory for the smaller contractor to ensure that this had been done by the main contractor. However, assurance must be sought.
		Co-operate and seek co-operation *5(1)*	**Co-operate and seek co-operation** *5(1)*	All duty holders must co-operate with and actively seek the co-operation of those involved in construction on the same or adjoining sites. **Note:** The project site is also an adjoining site to those around. This will remain a dynamic situation throughout the project.
		Co-ordination *6*	**Co-ordination** *6*	All site activities need due co-ordination in consideration of the health and safety of those carrying out construction work or affected by the carrying out of construction work. The concept of the main contractor can facilitate this process.
			General principles of prevention *7*	Such principles, which represent the basic tenets of health and safety management, must be applied in carrying out the construction work. This also reinforces the requirements of the MHSW Regulations 1999. All such work under the contractor's control must embrace the general principles of prevention. This also relates to remedial work and repairs after handover undertaken as part of the contractual arrangement.
			Plan, manage and monitor construction work *13(2)*	Continual monitoring is essential to ensure the effectiveness of the systems of control. This allows a proactive response to address situations in advance of failure.
			Mobilisation period *13(3)*	All contractors must inform those they engage or appoint of the minimum amount of time allowed for planning and preparation before they begin construction work.

Continued

Initial design phase	Significant detailed design phase	Pre-construction phase	Construction phase	Post-construction phase
	Note: All contractors must also manage their site activities in full compliance with the requirements of Part 4 of the CDM Regulations 2007 Duties Relating to Health and Safety on Construction Sites, i.e. Regulations 25 to 44 inclusive.	Such induction and information must be in a comprehensible form and reflect the ethnic mix on site.	**Information and training** 13(4)	All contractors must provide every worker under their control with any relevant information and training including: • site induction • risk assessment information • risks arising from other contractors' undertakings • workplace methodologies relating to the above • site rules • emergency procedures • identity of persons nominated to implement emergency procedures.
			Health and safety training 13(5)	This is a cross-reference to the MHSW Regulations 1999 and the need to provide necessary training to safely undertake the construction work.
		Construction sites are hazardous environments and thus measures must be taken to prohibit entry to unauthorised personnel from the start of construction.	**Unauthorised access** 13(6)	No construction work shall begin unless reasonable steps have been taken to prevent access by unauthorised persons to the site.
		Basic welfare facilities are essential in all workplaces but particularly on construction sites by the nature of work undertaken and the need to maintain and implement high standards of personal hygiene.	**Welfare** 13(7)	Adequate and suitable welfare facilities must be available from day one of construction and throughout the construction phase. Schedule 2 provides details of welfare facility requirements.

Figure 6.2
CONTRACTOR: ADDITIONAL DUTIES ON NOTIFIABLE PROJECTS

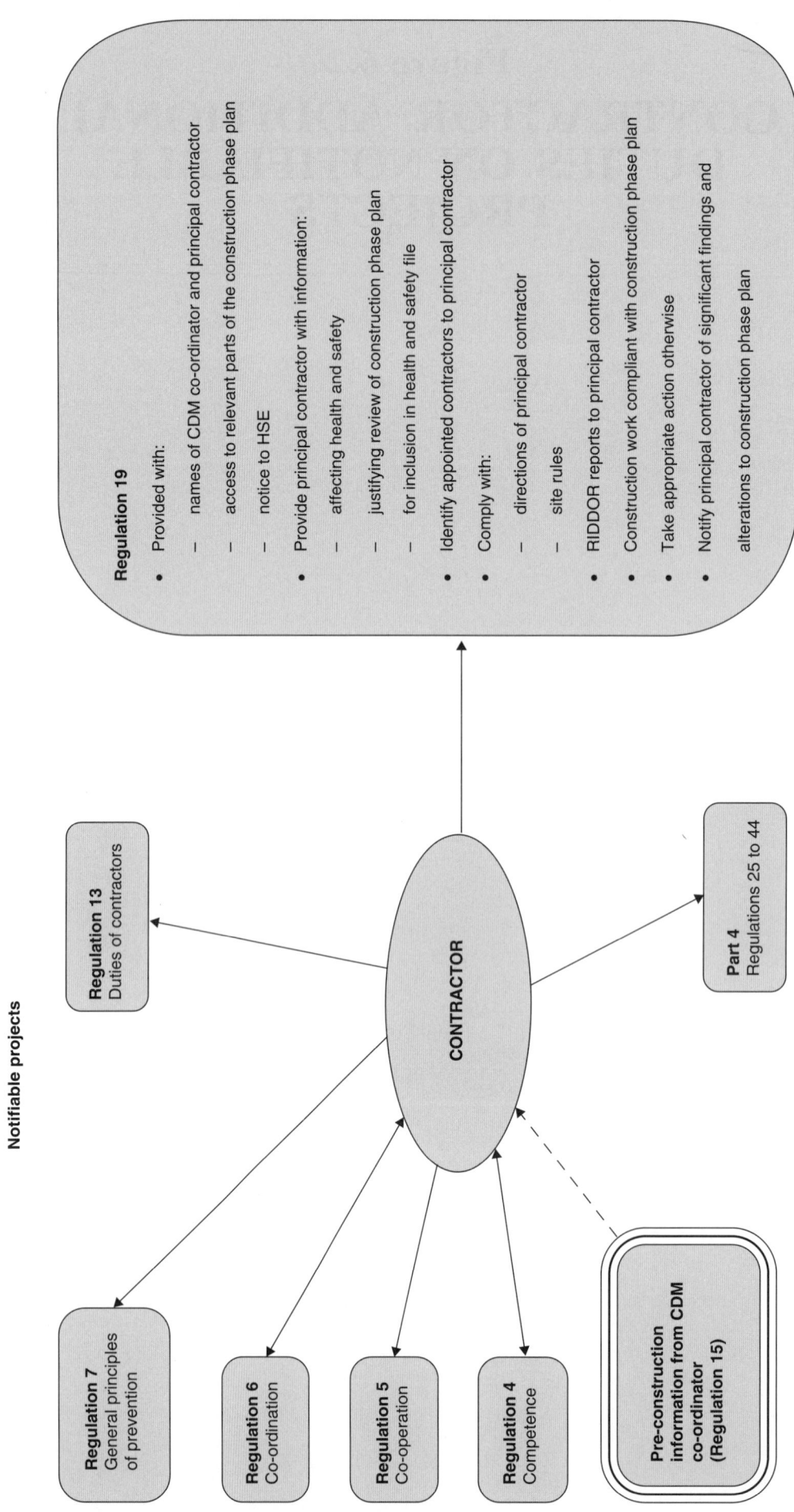

Figure 6.2   Contractor: additional duties on notifiable projects

6.2
CONTRACTOR FLOWCHART: NOTIFIABLE PROJECTS, ADDITIONAL DUTIES

Initial design phase	Significant detailed design phase	Pre-construction phase	Construction phase	Post-construction phase
		Actions to be confirmed *19(1)(a), (b) and (c)*	No construction work is to start unless the contractor is provided with: ● name of CDM co-ordinator and principal contractor ● access to relevant part of the construction phase plan ● evidence that notification has been given to the HSE or to the Office of Rail Regulation.	**Sufficient detail must be provided** This relates to work to be performed by the contractor
			Provide information to the principal contractor *19(2)(a)*	Information must be provided promptly to the principal contractor in respect of workplace risk assessments: ● affecting the health and safety of those carrying out construction work or affected by it ● justifying a review of the construction phase plan ● identified for inclusion in the health and safety file.
		This is to enable the principal contractor to maintain control over the entire process of site management.	**Identify all contractors directly appointed or engaged** *19(2)(b)*	
			Comply with directions *19(2)(c)*	Contractors must comply with reasonable directions given by the principal contractor and any site rules.
		The competent person for the contractor must also inform the HSE.	**RIDDOR** *19(2)(d)*	Principal contractor must be provided with information required under the Reporting of Injuries, Diseases and Dangerous Occurrences Regulations (RIDDOR) 1995.
		The construction phase plan should have been drafted by the principal contractor in consultation with each relevant contractor.	**Construction phase plan** *19(3)(a)*	Every contractor must ensure work is undertaken compatible with the construction phase plan.
			Appropriate action *19(3)(b)*	Proactive response required to ensure effective health and safety control where it is not possible to comply with the construction phase plan.
		This requires communication with the principal contractor so that the necessary review and amendment can be initiated.	**Notify principal contractor** *19(3)(c)*	Principal contractor to be notified of significant findings requiring alteration/amendment to the construction phase plan, preferably in writing so that a record of evidence exists.

6.3
CONTRACTOR CHECKLIST: ALL PROJECTS

CONTRACTOR CHECKLIST: ALL PROJECTS (Sheet 1 of 4)

Contract

No.	Reg.	Stage	Procedure	Description	Timing of action	Action req. Yes	Action req. No	Initials	Date actioned
1	4(1)(a)	Pre-appointment	**APPOINT COMPETENT DUTY HOLDERS**	All duty holders, i.e. designer(s) and contractor(s) appointed directly by the contractor, must be assessed as competent. The contractor must be satisfied that the person/organisation appointed has the necessary competence to perform the function. Reliance cannot solely be placed on the self-certification process (Regulation 4(1)(b) below). The rigour of the assessment must relate to the complexity of the project.	Before appointment and often well in advance via the selection procedures associated with the Approved List.				
2	4(1)(b)		**SELF-CERTIFICATION**	Acceptance of such an appointment signifies competence. Any party accepting a duty holder appointment self-certifies that they are a competent person.	Before appointment.				
3	13(1)	Pre-construction start	**ENSURE CLIENT AWARENESS**	All contractors must ensure the client is aware of his duties under these regulations. There are no exceptions. This can be facilitated by ensuring that the lead contractor (principal contractor or main contractor) has notified the client.	Before the start of construction work. Beyond this, the client's understanding can be gauged by his performance in discharging his duties. If this understanding is not apparent, then further action must be undertaken to correct the situation.				
4		Before the relevant stage of construction	**RECEIVE PRE-CONSTRUCTION INFORMATION**	Pre-construction information is to be issued promptly by the client to all contractors appointed or who might be appointed by the client.	To enable the contractor to effectively embrace relevant information within the development of his safe and suitable systems of work under his management of the construction process.				

Continued

Contract					Action req.			
No.	Reg.	Stage	Procedure	Description	Timing of action	Yes No	Initials	Date actioned
5 6 7	5(1)(a) 5(1)(b) 5(2)	All stages	**CO-OPERATE AND ACTIVELY SEEK THE CO-OPERATION REPORTING**	All duty holders must co-operate with all persons involved in construction work at the same or an adjoining site. Such co-operation must be actively sought. All duty holders must report anything likely to endanger themselves and/or others to those in control. **Note:** Your project is itself an adjoining site.	Particularly at the start of the project but actively throughout the project period. Throughout the construction period. Throughout the construction period.			
8		All stages	IDENTIFICATION	There is a continuing need to identify all those with whom the contractor needs to co-operate. This could well relate to overlapping projects, etc.	Particularly at the start of the project but actively throughout the project period.			
9	6	All stages	**CO-ORDINATE**	Contractors must co-ordinate their activities with other duty holders to ensure the health and safety of those carrying out and/or affected by such construction work.	Particularly at the start of the project but actively throughout the project period. This should be facilitated by the principal contractor.			
10		All stages	IDENTIFICATION	There is a continuing need to identify all those with whom the contractor needs to co-ordinate activities. This could well relate to overlapping projects, etc. Emphasis is placed on the role of the lead contractor (main contractor or principal contractor) in the process of co-ordination. Co-ordination, like co-operation and communication, is an essential component of effective project management.	Particularly at the start of the construction period but actively throughout the project.			

Continued

CONTRACTOR CHECKLIST: ALL PROJECTS (Sheet 2 of 4)

Contract								
No.	Reg.	Stage	Procedure	Description	Timing of action	Action req. Yes No	Initials	Date actioned
11	7	All stages	**GENERAL PRINCIPLES OF PREVENTION**	The basic tenets of health and safety management must be embraced by all duty holders. Such compliance also links with the management of construction duties, particularly under Regulation 13(4)(c).	Throughout the project period.			
12	*13(2)*		**PLAN, MANAGE AND MONITOR CONSTRUCTION WORK**	All contractors must ensure that safe and suitable systems of work remain as effective controls irrespective of the way in which construction work is carried out. Construction work must be carried out without risk to health and safety. Contractors must be able to demonstrate that such work was undertaken in compliance with the term *'as far as is reasonably practicable'*.	Throughout the construction process.			
13	*13(3)*		**INFORM ABOUT MOBILISATION PERIOD**	This is the minimum period for planning and preparation before construction work begins. It ensures that the contractor does not begin construction work prematurely and/or when ill-prepared. Ideally, agreement on this period of mobilisation should be made through dialogue between the parties. Such information should be forwarded to all those contractors appointed and subsequently received from all those who appoint.	Simultaneously with the appointment/engagement of those contractors directly appointed by the contractor.			

Continued

No.	Reg.	Stage	Procedure	Description	Timing of action	Action req. Yes	Action req. No	Initials	Date actioned
14	13(4)		**PROVIDING INFORMATION AND TRAINING**	As a basic requirement of health and safety law every worker should be given: • suitable site induction • relevant information in respect of risk assessments undertaken under the MHSW Regulations 1999 or arising out of the conduct of another contractor • control measures in respect of risk assessments • site rules • emergency procedures • identity of nominated person implementing emergency procedures.	At the beginning of construction and throughout as necessitated by the changing nature of construction within the framework of the construction programme. Such induction and information needs to be in a comprehensible form and must reflect the ethnic mix on site.				
15	13(5)		**PROVIDE HEALTH AND SAFETY TRAINING**	Reference is made to Regulation 13(2)(b) of the MHSW Regulations 1999, which requires every employer to ensure that his employees are provided with adequate health and safety training on their being exposed to new or increased risks.	As necessary throughout the construction as dictated by the nature of the work.				
16	13(6)		**PREVENT UNAUTHORISED ENTRY**	The site must be secure from entry by unauthorised persons. This reinforces the duties laid down in the Occupiers' Liability Act 1984 with a duty owed to the trespasser.	From day one and throughout the construction phase.				
17	13(7)		**PROVIDE WELFARE PROVISION**	Welfare requirements as detailed in Schedule 2 must be provided as far as is reasonably practicable throughout the construction phase.	Adequate and suitable welfare facilities must be provided from day one and remain effective throughout the construction phase.				

CONTRACTOR CHECKLIST: ALL PROJECTS (Sheet 4 of 4)

6.4
CONTRACTOR CHECKLIST: NOTIFIABLE PROJECTS, ADDITIONAL DUTIES

No.	Reg.	Stage	Procedure	Description	Timing of action	Action req. Yes	No	Initials	Date actioned
A	19(1)(a) 19(1)(b) 19(1)(c)	Pre-construction	RECEIVE INFORMATION CONCERNING: • NAME OF CDM CO-ORDINATOR AND PRINCIPAL CONTRACTOR • RELEVANT SECTIONS OF CONSTRUCTION PHASE PLAN WITH SUFFICIENT DETAIL IN RESPECT OF WORK TO BE PERFORMED • NOTIFICATION TO THE HSE OR THE OFFICE OF RAIL REGULATION.	Most of this information would be contained on the F10 notice as displayed and also in the related information passed over by the principal contractor in the construction phase health and safety plan.	Prior to starting any construction work.				
B	19(2)(a)	Construction	PROVIDE RELEVANT INFORMATION, INCLUDING ANY RISK ASSESSMENT INFORMATION, TO THE PRINCIPAL CONTRACTOR WHICH MIGHT AFFECT: • HEALTH AND SAFETY OF ANYBODY CARRYING OUT CONSTRUCTION WORK OR AFFECTED BY IT • A REVIEW OF THE CONSTRUCTION PHASE PLAN • IDENTIFIED FOR INCLUSION IN THE HEALTH AND SAFETY FILE.	Such information is to be provided promptly.	Throughout the contract.				
C	19(2)(b)	Construction	IDENTIFY TO THE PRINCIPAL CONTRACTOR ANY CONTRACTOR DIRECTLY APPOINTED OR ENGAGED.	This facilitates the principal contractor's ability to manage the project in its entirety.	Immediately on appointment and throughout the construction phase.				
D	19(2)(c)	Construction	COMPLY WITH BOTH DIRECTIONS AND SITE RULES FROM PRINCIPAL CONTRACTOR.	The principal contractor co-ordinates health and safety management on site and all contractors must comply with reasonable instructions given.	Throughout the contract.				
E	19(2)(d)	Construction	INFORMATION TO BE PROVIDED TO THE PRINCIPAL CONTRACTOR IN RESPECT OF RIDDOR REPORTING PROCEDURES.	Such reports must also be made by the competent person for the contractor to the HSE as well as to the principal contractor.	On the occurrence of death, injury, disease or dangerous occurrence as required by the RIDDOR 1995.				
F	19(3)(a) 19(3)(b) 19(3)(c)	Construction	ENSURE THE CONSTRUCTION PHASE PLAN: • IS COMPLIED WITH • IS ALTERED OR ADDED TO IF FOUND DEFICIENT. APPROPRIATE HEALTH AND SAFETY MANAGEMENT ACTION TO BE TAKEN IF CONSTRUCTION PHASE PLAN IS INEFFECTIVE.	Reasonable steps to be taken to ensure that this is done. Proactivity and monitoring are essential prerequisites to ensure that health and safety are effectively delivered.	Throughout the contract.				

Section 7
THE CDM
CO-ORDINATOR

7.1 Introduction

The CDM co-ordinator function arises directly as a 'creature' of the Regulations and is pivotal to the entire communication and co-ordination process underlying the implementation of the regulations on the notifiable project.

As noted in the ACoP:

'The role of the CDM co-ordinator is to provide the client with a key project advisor in respect of construction health and safety risk management matters. They should assist and advise the client on appointment of competent contractors and the adequacy of management arrangements; ensure proper co-ordination of the health and safety aspects of the design process; facilitate good communication and co-operation between project team members and prepare the health and safety file.'

This accords far more closely with the original directive in both name and function than the earlier role of the planning supervisor (CDM Regulations 1994) but does not translate into being the:

- technical auditor or
- safety advisor.

Contractually, the CDM co-ordinator has limited powers but functions from a position of strength in the implementation of criminal legislation and as key advisor to the client. In reality, the function is greatly facilitated by the visible support of the client both in establishing the necessary protocols and through the prompt discharge of his own duties.

7.2 Appointment

This role is dependent on the CDM co-ordinator being appointed by the client in writing as soon as is practicable after the initial design phase. It is only the client who can appoint.

Failure to appoint at the requisite time introduces:

- limitations to the role
- the default mechanism whereby the client takes on the duties
- compromise in the function of those designing and those constructing.

Whilst the role of CDM co-ordinator is aligned to that of the former planning supervisor, it should be noted that duties have been altered and the competence and skill-set of the planning supervisor needs extension into areas of delivery that now reside with the CDM co-ordinator.

The CDM co-ordinator's role is to facilitate the delivery of health and safety risk management within the construction project environment. It consists of support primarily to the client, but also to other duty holders in ensuring that arrangements are made and implemented for the co-ordination of health and safety measures at the appropriate time. It is better seen as a team-based function, although it could be delivered by the sole practitioner, partnership or limited company. The team-based perspective allows inputs from additional team members to account for the specific insights required on the wide variety of projects across the construction spectrum.

This support is fundamental to the success of any notifiable construction project and should be given as early as possible. The CDM co-ordinator needs to be aware of the additional resources and competence thus required.

The law also allows the same party to function in all duty holder roles subject to considerations of competence and adequacy of resource. However, prudence suggests that management systems need further development and additional sophistication, with the need for impartiality and objectivity tantamount to successful delivery in each role.

There is no transitional period and as Regulation 47(4) notes:

'Any planning supervisor or principal contractor appointed under Regulation 6 of the 1994 Regulations shall, in the absence of an express appointment by the client, be treated for the purpose of paragraph 2 as having been appointed as the CDM co-ordinator, or the principal contractor, respectively.'

Regulation 47(5) then goes on to remind us that such appointments have 12 months from the enactment of these regulations (6 April 2007) for steps to be taken to ensure competence is compatible with the requirements of Regulation 4(2), i.e. *'to perform any requirements and avoid contravening any prohibition, imposed on him by or under any of the relevant statutory provisions.'* **In all other cases the appointment must be based on the establishment of competence at the time of appointment.**

Failure to appoint at the appropriate time ensures that the client takes on such responsibilities by default (refer to Regulation 14(4)).

The appropriate time is not well defined. The ACoP (paragraphs 66 and 86) draws attention to early appointment, with the appointment taking place 'as soon as practicable after initial design work or other preparation for construction work has begun' and distinguishes between initial design and significant detailed design work, which includes preparation of the initial concept design and implementation of any strategic brief. Misinterpretation of the design stage for appointment will again ensure that the client takes on these duties by default.

The wisdom is to appoint sooner rather than later to avoid the chasm of misinterpretation.

The CDM co-ordinator has been promoted as the client's best friend, and certainly with the increase in client duties, support and advice are required. Hence the relationship between the client and his CDM co-ordinator is critical for the successful delivery of effective health and safety risk management on construction projects to ensure that **the right information gets to the right people at the right time**.

Figure 7.1
CDM CO-ORDINATION DUTIES

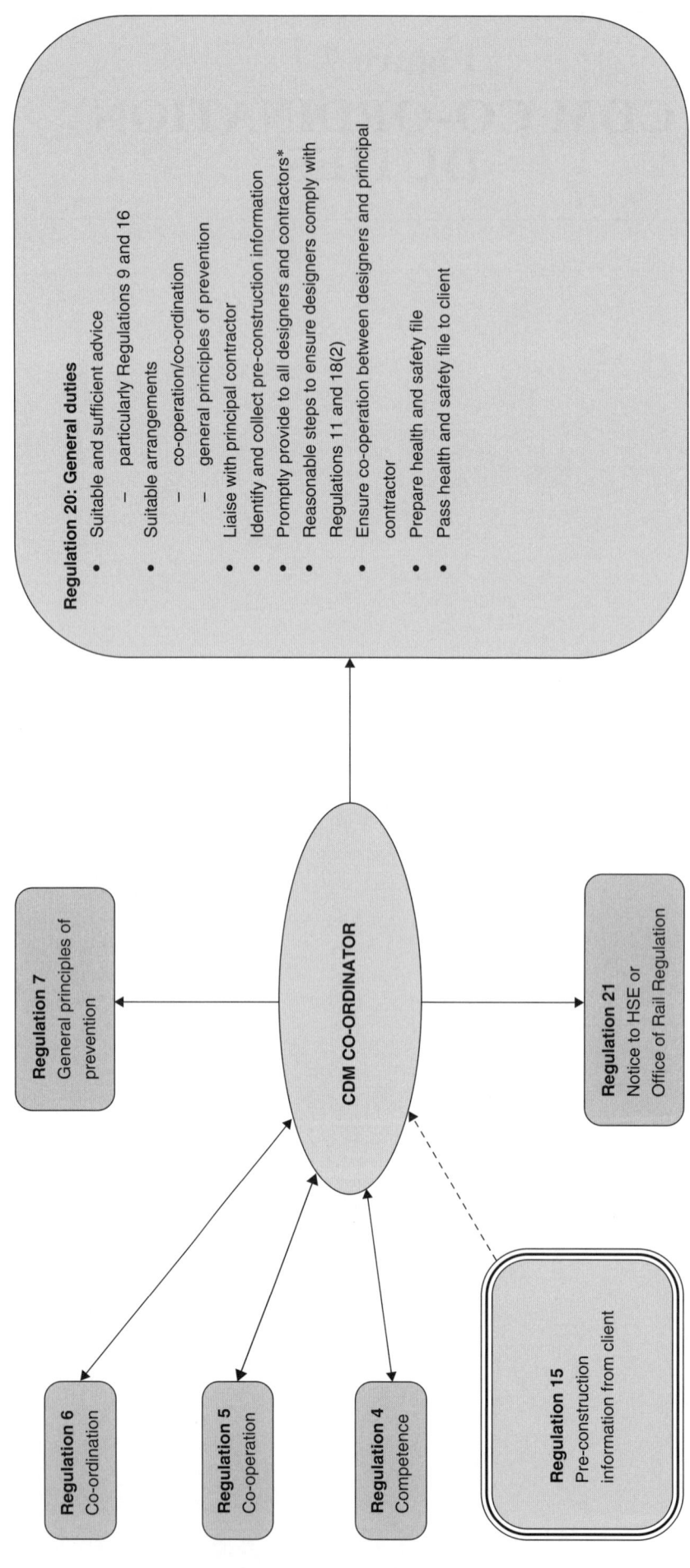

Figure 7.1 CDM co-ordination duties
*Only those contrators who have been, or who might be, appointed by the client.

7.1
CDM CO-ORDINATOR FLOWCHART

Initial design phase	Significant detailed design phase	Pre-construction phase	Construction phase	Post-construction phase
Appoint competent duty holders *4(1)*	All duty holders whenever appointed must be assessed for competence. In accepting, the duty holder self-certifies that they are competent.	No person appointing can rely on self-certification arising out of Regulation 4(1)(b)	**The appointment of the CDM Co-ordinator (Regulation 14(1)) must be made in writing by the client (Regulation 14(5)).**	
This is a critical interface in the client's undertaking of his duties, with particular emphasis placed on compliance with Regulations 9 and 16.	**Give suitable and sufficient advice and assistance to the client** *20(1)(a)*	**Give suitable and sufficient advice and assistance to the client** *20(1)(a)*	**Give suitable and sufficient advice and assistance to the client** *20(1)(a)*	The CDM co-ordinator is therefore particularly involved in ensuring that management arrangements are effective throughout and that due control is exercised over the start of construction.
Notification should be sent off as soon as practicable after the appointment of the CDM co-ordinator. Notification either goes to the HSE or to the Office of Rail Regulation.	**Notification of project** *(initial)* 21	**Notification of project** *(additional)* 21	Additional information is sent when available, e.g. name and details of principal contractor. Copy to be provided to principal contractor and client.	It is imperative that this notice is signed by the client or someone on his behalf as endorsement that the client is aware of his duties under these regulations.
	Pre-construction information *15(a)*	**Pre-construction information** *15(a)*	Now that the CDM co-ordinator has been appointed, pre-construction information is passed from the client to the CDM co-ordinator for distribution.	**Note:** Such information might need delivery into the construction phase.
	Suitable arrangements *20(1)(b)*	**Suitable arrangements** *20(1)(b)*	These are for the co-ordination of health and safety measures during planning and preparation for the construction phase.	As such they include the facilitation of co-operation and co-ordination as well as the general principles of prevention, namely Regulations 5, 6 and 7.
		Such liaison could be initiated earlier dependent on procurement strategies. Early involvement of the principal contractor provides an opportune perspective on constructability, etc.	**Liaise with the principal contractor** *20(1)(c)*	Such liaison is for purposes of: ● health and safety file contents ● information needed for preparation of construction phase plan ● design developments affecting planning/ managing construction work.
	Identify and collect the pre-construction information *20(2)(a)*	**Identify and collect the pre-construction information** *20(2)(a)*	It is important that such identification is undertaken in dialogue with the design teams and other consultants employed by the client.	It would be imprudent for unilateral identification of information to be made by the CDM co-ordinator to the exclusion of other team members.
Timely delivery is essential and relevant pre-construction information should go to every person designing the structure and every contractor who has or may be appointed by the client.	**Promptly provide pre-construction information** *20(2)(b)*	**Promptly provide pre-construction information** *20(2)(b)*	**Promptly provide pre-construction information** *20(2)(b)*	This also includes delivery to the principal contractor and relates to information that is relevant to the role of principal contractor. The flow of relevant information is not limited to the pre-construction phase.

Continued

Initial design phase	Significant detailed design phase	Pre-construction phase	Construction phase	Post-construction phase
	Ensure designers comply with their duties *20(2)(c)*	**Ensure designers comply with their duties** *20(2)(c)*	All reasonable steps must be taken to ensure designers comply with duties under Regulations 11 and 18(2), which relate to design duties generally and the sufficiency of information about aspects of the design of the structure or its construction or maintenance as will assist the CDM co-ordinator in complying with his duties, including duties in respect of the health and safety file.	Reasonable steps are required to ensure that design is considering and contributing to health and safety management and relevant information is being delivered to those who need it. The CDM co-ordinator needs to be integrated into the design process for this to be effectively achieved. Consideration could be given to attendance at design review meetings, etc.
			Co-operation between designers and the principal contractor *20(2)(d)*	Reasonable steps are to be taken by the CDM co-ordinator to ensure co-operation between these two parties in respect of design development and/or design change. As with other duties this requires a proactive approach by the CDM Co-ordination team.
	Note: The preparation of the health and safety file starts at the same time as the appointment of the CDM co-ordinator or shortly after. Questions of format, multiple copies and compatibility of client document management systems must all be addressed early on in the process, as well as the eventual provider of the 'as-built' record drawings.	**Prepare the health and safety file** *20(2)(e)*	**Prepare the health and safety file** *20(2)(e)*	This positions the CDM co-ordinator with full responsibility for this document. Much support can be provided by the client, but failure to produce such a document exposes the CDM co-ordinator to charges of non-compliance. This duty links in with the provision of information from clients (Regulation 10(2)) and designers (Regulation 18(2)) and the contractor's duty (Regulation 19(2)(a) under the umbrella of the principal contractor's duty to identify such information (Regulation 22(1)(j)).
	The health and safety file must be passed to the client at the end of the construction phase. Timing is critical and it should coincide with the insurance switch for the project back to the client from the principal contractor.	**Note:** Partial handovers and sectional completions require a relevant health and safety file to be handed over at the requisite time.	All handovers should be accompanied by a letter explaining the nature and purpose of the health and safety file and a signature of receipt obtained.	**Pass the health and safety file to the client** *20(2)(f)*

7.2
CDM CO-ORDINATOR CHECKLIST

The following procedures are requirements for all notifiable projects (a CDM co-ordinator would not be required on a non-notifiable project).

| Contract | | | | | | | Action req. | | Initials | Date actioned |
No.	Reg.	Stage	Procedure	Description	Timing of action		Yes	No		
A	14(5)	Post initial design stage. Before the start of significant detailed design	**NOTIFICATION OF APPOINTMENT**	**The appointment of the CDM co-ordinator must be made in writing by the client** It would be imprudent to fulfil this role without the written confirmation of appointment. Where the role of CDM co-ordinator has been fulfilled within the client organisation either by design and/or default care needs to be exercised and conditions/qualifications made about the service being subsequently provided.	Before significant detailed design. **Note: Failure to appoint a CDM co-ordinator at the appropriate time ensures that the client fulfils the role.** **There are no belated appointments in respect of CDM co-ordination since failure to appoint ensures that the client takes on such duties.**				
B	20(1)(a)	On appointment	**GIVE SUITABLE ADVICE AND ASISTANCE**	This is a critical interface in the client's undertaking of his duties, with particular emphasis placed on compliance with Regulations 9 and 16, management arrangements and the control over the start of construction, respectively.	Throughout the project, but with particular reference at the beginning.				
C		Design/construction	**MANAGEMENT ARRANGEMENTS**	Regulation 9 requires that the client's obligations must ensure that management arrangements are suitable and remain effective throughout the project. The CDM co-ordinator must provide guidance in this area.						
D			**CONFIRM SUFFICIENCY OF CONSTRUCTION PHASE PLAN TO CLIENT**	The role of the CDM co-ordinator in the facilitation of control over the start of construction is a key area of advice, with the CDM co-ordinator advising the client on the sufficiency of the principal contractor's construction phase plan as a condition precedent to starting the construction phase. However, it must be noted that it is the client who ultimately sanctions the start of construction, albeit on the advice of his CDM co-ordinator	Pre-construction phase start inclusive of any preparatory works.				

Continued

No.	Reg.	Stage	Procedure	Description	Timing of action	Action req. Yes	Action req. No	Initials	Date actioned
				This document is either sufficient or not and hence a conditional start **should not** be given for the construction phase plan needs to be adequately developed to ensure that **safe and suitable** systems of work have been detailed and articulated by the principal contractor in his control of all early site activities. The client's understanding of his duties can be gauged by performance. This performance itself needs to be monitored and appropriate action taken if found wanting. **A PROJECT THAT DOES NOT START WITH EFFECTIVE CONTROLS IS UNLIKELY TO DELIVER SUCCESSFUL HEALTH AND SAFETY MANAGEMENT.**					
E	21(1)	Post initial design/pre significant design	**ENSURE INITIAL F10 NOTIFICATION TO EITHER THE HSE OR THE OFFICE OF RAIL REGULATION**	This must be sent to the HSE or the Office of Rail Regulation giving relevant information about the project as outlined in Schedule 1. HSE form F10 is normally used for this purpose. **Ensure client or his representative has signed the F10 to confirm that he understands his duties with respect to the CDM Regulations.** **It is only the client or his representative who now has to sign the F10 notification.**	As soon as practicable after appointment. Notification should go to the regional HSE office associated with the location of the construction site.				
F			COPY TO BE FORWARDED TO THE CLIENT	For client's record purposes.					

Continued

Contract

No.	Reg.	Stage	Procedure	Description	Timing of action	Action req. Yes	Action req. No	Initials	Date actioned
G	15(a) and 15(b)		**RECEIVE PRE-CONSTRUCTION INFORMATION**	The client is under an obligation to pass relevant pre-construction information to the CDM co-ordinator. Such information is to be provided promptly for distribution by the CDM co-ordinator to: • all designers, and • contractors who have been or who may be appointed by the client. Such information relates to: • that about or affecting the site • the proposed use of the structure • mobilisation period • existing health and safety file information. **In keeping with the concept of information flow this becomes a project dynamic, with relevant additional information being distributed as it arises.**	As soon as practicable after appointment and forwarded at the appropriate time. This links in with the concept of information flow and will therefore be distributed to relevant parties compatible with the development stages of each project. The process will need to identify what information goes to which duty holder at the appropriate time, e.g. **pre-construction information: tender stage** will need to be distributed to all contractors tendering to the client at the point of forwarding the tender documentation.				
H	20(1)(b)		**ENSURE SUITABLE ARRANGEMENTS FOR CO-ORDINATION AND CO-OPERATION**	Co-operation and co-ordination remain essentials for project management success. The CDM co-ordinator must be satisfied that suitable arrangements are in place for co-operation and co-ordination between all duty holders, particularly during the planning and management stages leading up to the start of the construction phase, but also that such arrangements remain effective thereafter. Such assurance can be obtained via confirmation received through appropriate agenda items. Additionally, satisfaction must be obtained that all duty holders are compliant with the general principles of prevention throughout their processes.	Before the start of the construction phase and throughout the construction phase. As a project dynamic co-operation and co-ordination must embrace all relevant parties at the appropriate time. The list of those with whom co-ordination and co-operation must take place will change throughout the project. This list needs constant revision.				

Continued

No.	Reg.	Stage	Procedure	Description	Timing of action	Action req. Yes	Action req. No	Initials	Date actioned
J	20(2)(a)		**IDENTIFY AND COLLECT PRE-CONSTRUCTION INFORMATION**	This is the pre-construction information required by various duty holders in order to discharge their statutory duties. Such advice and assistance would reflect on the status of the client, with seemingly more guidance needed for the lay client than for the informed client. Care should be exercised that this is not done unilaterally but in conjunction with the design team and other professional advisors.	As soon as practicable after appointment and throughout the process thereafter.				
K	20(2)(b)		**PROMPTLY PROVIDE PRE-CONSTRUCTION INFORMATION**	The timing is important and must be at a time that allows the duty holders to embrace the information for the purpose of effective health and safety management. Such information as is relevant needs to be provided to: • all designers and • contractors who have been or who may be appointed by the client. Such duty holders need to be identified as soon as possible and include specialist package designers appointed late into the construction phase period.	This also includes delivery to the principal contractor after appointment, the extent of which will depend on the procurement strategy. This underlines the fact that this distribution is not limited to the pre-construction phase.				
L	21(2)	Pre-construction phase	**ENSURE ADDITIONAL F10 NOTIFICATION TO EITHER THE HSE OR THE OFFICE OF RAIL REGULATION**	**Additional relevant information must be sent to either the HSE or the Office of Rail Regulation in respect of particulars not known at the time of the initial notification. This usually would relate to details about the principal contractor.**	As soon as practicable after appointment of principal contractor.				

Continued

Contract								
No.	Reg.	Stage	Procedure	Description	Timing of action	Action req. Yes	No	Initials	Date actioned
M			COPY TO BE FORWARDED TO THE CLIENT AND PRINCIPAL CONTRACTOR	The client should already have signed the initial notification. The principal contractor must display this on site. Amendments are required to such information if significant change occurs, e.g. the appointment of a new principal contractor or if the start date changes by a month or more.	Prior to start of construction.				
N	*20(1)(c)*		**LIAISE WITH PRINCIPAL CONTRACTOR**	Early liaison provides the principal contractor with the optimum period for identifying contractor input to the health and safety file. Such liaison is for the purposes of: • health and safety file contents • information needed for preparation of the construction phase plan • design developments affecting planning/ managing construction work.	Before the start of construction and throughout the design and construction phases. Design change during the confines of a construction programme has a greater impact on the health and safety controls being exercised, and should be assessed vigilantly.				
O	*20(2)(c)*		**ENSURE DESIGNERS COMPLY WITH THEIR DUTIES**	The CDM co-ordinator must be satisfied that designers are fulfilling their duties with respect to Regulations 11 and 18(2). This can be achieved via: • discussions and dialogue • attendance at design review and development meetings • correspondence exchange • contributions to the project risk register.	Throughout the design phase.				

Continued

No.	Reg.	Stage	Procedure	Description	Timing of action	Action req. Yes	No	Initials	Date actioned
P	20(2)(d)		**ENSURE CO-OPERATION BETWEEN DESIGNERS AND THE PRINCIPAL CONTRACTOR**	This interface is particularly critical because of the confines imposed by the construction programme. There is a major implication in respect of late design changes and their effect on health and safety management as well as the liaison demanded between permanent and temporary design.	Throughout the construction phase.				
Q	20(2)(e)		**PREPARE THE HEALTH AND SAFETY FILE**	The CDM co-ordinator is liable for the preparation of the new health and safety file or the review and update of an existing health and safety file. The non-appearance of such a document exposes the CDM co-ordinator to a charge of non-compliance. The management of health and safety file information commences on the appointment of the CDM co-ordinator and continues throughout the subsequent process. Arrangements must be set up to identify and receive such information as and when it becomes available. This does not mean that the CDM co-ordinator must physically prepare – it can still be collated by a third party, possibly the principal contractor by agreement. However, the process must be continually managed by the CDM co-ordinator, who remains liable for the preparation of such a document.	Throughout the design/construction process.				

Continued

CDM CO-ORDINATOR CHECKLIST (Sheet 6 of 7)

Contract

No.	Reg.	Stage	Procedure	Description	Timing of action	Action req. Yes	No	Initials	Date actioned
R	20(2)(f)		**PASS THE HEALTH AND SAFETY FILE TO THE CLIENT**	The health and safety file should be passed to the client at the end of the construction phase.	In conjunction with the handover of the project.				
				This handover should coincide with the project insurance changeover from principal contractor to client.					
				Sectional completion/partial handover should be similarly accompanied by the appropriate health and safety file.					
S			EXPLANATORY LETTER	The handover should be accompanied by an explanatory letter to the client outlining the purpose of the health and safety file and the future obligations of the client in respect of revision, update and handover with change of ownership.					
				Since it is another key document it should be delivered personally or sent by recorded delivery.					
T			SIGNATURE ON RECEIPT	A record of handing the health and safety file over to the client should be obtained.	This completes the CDM co-ordination role and concludes the administrative obligations for the project.				

Section 8
THE PRINCIPAL CONTRACTOR

As noted in paragraph 146 of the ACoP:

'The key duty of principal contractors is to properly plan, manage and co-ordinate work during the construction phase in order to ensure that the risks are properly controlled.'

Principal contractors are responsible for managing and co-ordinating all construction phase health and safety issues without embarking on detailed supervision of the contractor's work. CDM provides a framework for this process, with the key risk management issues being set out in the construction phase plan.

Good management of health and safety is crucial to the successful delivery of a construction project and an ineffective approach in the execution of early supply chain duties has been identified as contributing to the 'accident on site' (refer to Research Report 218, *Peer Review of Analysis of Specialist Group Reports on Causes of Construction Accidents*, HMSO, 2004 (ISBN 0 7176 2836 1).

Under the CDM Regulations 2007 greater integration is required between the principal contractor and the CDM co-ordinator, as well as with his contractors (sub-contractors). Compliance with Regulation 22(1)(a), which requires the usual focus on planning, managing and monitoring the construction phase, together with the need to co-operate and co-ordinate as outlined for every duty holder in Regulations 5 and 6 endorses the concept of the integrated team in the delivery of effective health and safety management.

Further liaison is required with the CDM co-ordinator and designers in co-operating with matters of design and/or design change (Regulation 22(1)(b)) and additional dialogue promoted through consultation with contractors for the purposes of finalising the construction phase plan (Regulation 22(1)(g)) as well as the identification of inputs to the health and safety file by contractors (Regulation 22(1)(j)). Ownership of health and safety through consensus and contribution receives additional direction from the principal contractor's duty to consult the workforce on arrangements made and their effectiveness, as covered in Regulation 24.

The control process emphasises the need to monitor and liaise throughout, as well as manage the processes of construction. Control systems that are set up must continue to deliver the nurturing environment for all operatives and the words 'review, revise and refine' now qualify the management of the construction phase plan.

As well as the provision of welfare facilities compatible with the requirements of Schedule 2 from day one and throughout the contract, the principal contractor must inform contractors of the minimum amount of time for mobilisation and facilitate the information flow to contractors.

It is only the client who can appoint the principal contractor and failure to do so ensures that the client takes on these responsibilities (Regulation 14(4)(a)) by default. There can only be one principal contractor on the project and where overlaps occur between separate projects, communication, co-operation and co-ordination responsibilities require close attention.

Part 4 of the regulations sets out the duties associated with health and safety on construction sites, the response to which in respect of safe and suitable systems of work is articulated in the development of the principal contractor's construction phase plan, which must be sufficiently developed before any construction work starts.

The liaison required between the principal contractor and the CDM co-ordinator does not extend to the latter having a role in the development of the construction phase plan once construction has started, although the latter is pivotal to the client's sanction up to the start of the construction phase.

Generic documentation is not helpful in this respect and, as with all duty holders, project specific information must be provided. The construction phase plan is the principal contractor's articulation of his safe and suitable systems of work directly focused on duties relating to Part 4, Regulations 25 to 44, of the CDM Regulations 2007, as well as other construction-related legislation. Whilst due attention must be paid to its early development up to the sufficiency required to start construction, it remains imperative that this document is continually developed to be compatible with the programme of works.

It is a site document and demands contributions from duty holders (client, designers and CDM co-ordinator) before and throughout the construction phase as well as all contractors employed on site.

Successful delivery of any project, including health and safety, is dependent on the quality of the planning process in its entirety before work starts. Investment in this process lays the foundations for effective outcomes. The premature, ill-prepared project does not provide the controls or mitigation options to enable evasive action to be taken when the intended differs from the reality. It is the constant monitoring and review of systems that allows fine-tuning and modification to achieve the desired outcomes. Situations on site constantly change by the very nature of construction, and alertness and response are key parameters to maintaining the environment under which work can be undertaken safely.

Often the proactive process of planning and managing is thwarted by change generated by others and whilst the construction process must be flexible to accommodate change, such deviations challenge established controls. Many changes arise due to the lack of definition at stages within the project, as identified in Research Report 218. They also ensure that responses to such changes become disproportionately expensive. The minimisation of late changes rests with the project team and the concept of design as complete should be promoted by the project manager as soon as possible to be compatible with the procurement strategy, for the enhancement of project control generally and health and safety management specifically.

It is to be noted, as in paragraph 146 of the ACoP, that:

'Principal contractors must also comply with the duties placed on all contractors under the Regulations.'

Figure 8.1
REGULATIONS TO BE DISCHARGED BY PRINCIPAL CONTRACTOR

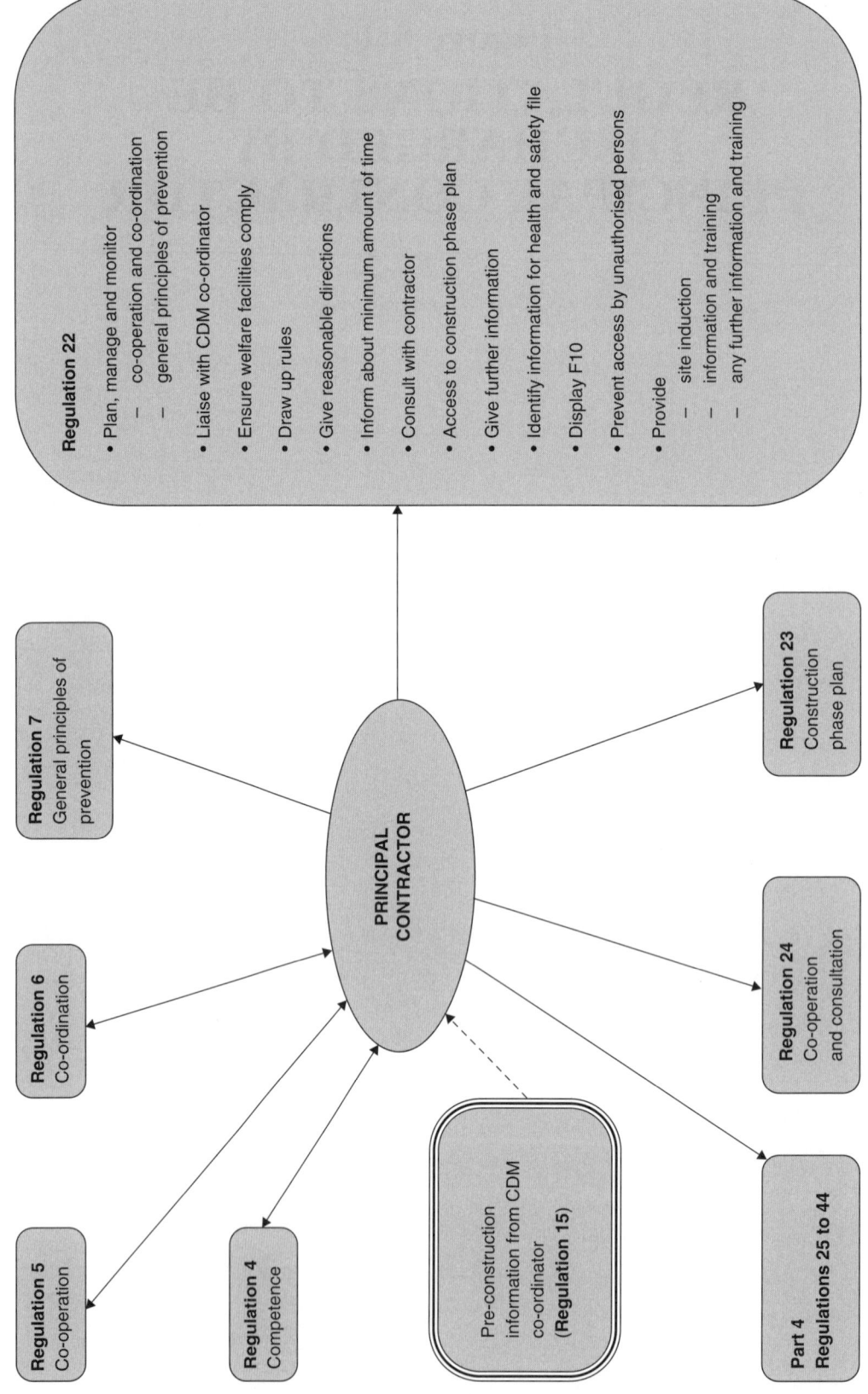

Figure 8.1 Regulations to be discharged by principal contractor

8.1
PRINCIPAL CONTRACTOR FLOWCHART

Initial design phase	Significant detailed design phase	Pre-construction phase	Construction phase	Post-construction phase
		Appoint competent duty holders *4(1)*	**Appoint competent duty holders** *4(1)*	All duty holders whenever appointed must be assessed for competence. In accepting such an appointment the duty holder self-certifies their competence. No person appointing can rely on self-certification arising out of Regulation 4(1)(b).
	As information flow updates may be required at different stages throughout the process depending on the procurement strategy.	RECEIVE PRE-CONSTRUCTION INFORMATION *20(2)(b)*	RECEIVE PRE-CONSTRUCTION INFORMATION *20(2)(b)*	The principal contractor must receive pre-construction information from the CDM co-ordinator. This information is essential for the development of the construction phase plan.
Note: The principal contractor must ensure that his tender submission accounts for the adequacy of relevant health and safety management resources in respect of day-to-day management issues that a competent principal contractor would expect to deal with and the specific site information contained within the pre-construction information tender stage document.				
		Prepare a construction phase plan *23(1)(a)*	This document must be prepared to a sufficient state before construction starts. Its sufficiency must ensure that all necessary management arrangements are in place so that from day one all matters of health, safety and welfare specific to the site are effectively under suitable management control. Reference is made to information from designers and pre-construction information from the CDM co-ordinator as essential prerequisites for the principal contractor in the development of this document into a sufficient state.	The sufficiency of the construction phase plan needs to be sanctioned by the client (Regulation 16(a)) before any construction work starts. There is a reminder within this sufficiency that the early stages of familiarisation with any project are the most critical. Each site remains hostile territory.
Note: All principal contractors must also comply with the duties placed on all contractors under the regulations.				
		This must be displayed throughout the construction period. This notification could have gone either to the HSE or the Office of Rail Regulation, depending on the nature of the work.	**Display notification to the HSE** *22(1)(k)*	Commonly referred to as the F10, this should be displayed in a readable condition in a position where it can be read by any worker engaged in the construction work. The usual place for display is the principal contractor's site office
Note: All principal contractors must also manage their site activities in full compliance with the requirements of Part 4 of the CDM Regulations 2007: Duties Relating to Health and Safety on Construction Sites, i.e. Regulations 25 to 44 inclusive.		Emphasis is placed on this document being a dynamic document and always in a state of development.	**Review, revise and refine the construction phase plan** *23(1)(b)*	The sufficiency of this document must continue compatible with the programme of works. Arrangements must be constantly monitored to ensure effectiveness and must be further developed in compliance with the changing pace of construction.

Continued

Initial design phase	Significant detailed design phase	Pre-construction phase	Construction phase	Post-construction phase
			Implement the construction phase plan 23(1)(c)	The implementation of the measures contained within the construction phase plan needs to embrace all persons who may be affected by the work. It therefore has a wider remit than focusing only on those who are employed.
			Take reasonable steps to ensure the construction phase plan identifies risks and control measures in respect of health and safety 23(2)	Constant vigilance is required to ensure that all relevant health and safety risks are identified and controlled. Failure to identify ensures lack of control. This process must be ongoing throughout the construction phase and must include associated work that runs into the maintenance period (snagging list).
		All duty holders must co-operate with and actively seek the co-operation of those involved in construction on the same or adjoining sites.	**Co-operate and seek co-operation** 5(1)	**Note:** The project site is also an adjoining site to those around. This will remain a project dynamic throughout the project.
		The principal contractor is the key co-ordinator in respect of site activities and control.	**Co-ordination** 6	All site activities need co-ordination in consideration of the health and safety of those carrying out construction work or affected by the carrying out of construction work.
		This also reinforces the requirements of the MHSW Regulations 1999.	**General principles of prevention** 7	Such principles, which represent the basic tenets of health and safety management, must be applied in carrying out the construction work.
		The principal contractor must ensure the effectiveness of these arrangements throughout the construction phase.	**Plan, manage and monitor construction phase** 22(1)(a)	Constant monitoring is essential to ensure planning and management arrangements remain effective. Reference is also made to this process facilitating co-operation, co-ordination and the application of general principles of prevention.
			Liaise with CDM co-ordinator 22(1)(b)	Close links are required with the CDM co-ordinator and designers in respect of design or changes to design during the construction phase.
			Ensure welfare facilities are sufficient 22(1)(c)	Adequacy of welfare compatible with the requirements of Schedule 2 must be provided on site from day one of the construction phase.
		Such rules and related information should be in comprehensible language and reflect the ethnic mix on site.	**Draw up site rules** 22(1)(d)	These are rules appropriate to the construction site and the activities on it. They must therefore be site specific.

Continued

Initial design phase	Significant detailed design phase	Pre-construction phase	Construction phase	Post-construction phase
			Give reasonable directions *22(1)(e)*	The directing influence on site is the role fulfilled by the principal contractor. This direction extends to all on the construction site throughout the period.
			Inform every contractor of the minimum period of mobilisation *22(1)(f)*	It is a minimum period and is that allowed for planning and preparation before the contractor begins construction work. Ideally the period should be ratified by dialogue between the principal contractor and the contractor.
			Consult contractors before finalising the construction phase plan *22(1)(g)*	Necessary engagement with the contractor ensures a sensible approach to the finalisation of the relevant parts of the construction phase plan.
		Such access must be given before such work is begun.	**Access to the construction phase plan** *22(1)(h)*	Every contractor must be allowed access to the relevant parts of the construction phase plan in sufficient time to enable him/her to prepare properly for the associated work.
		Induction processes and 'tool-box' talks are the usual vehicle for imparting such information.	**Further information** *22(1)(i)*	Such information is required to enable compliance punctually with the requirements of Schedule 2 (welfare arrangements) and to enable the related work to be carried out safely.
		These inputs must be identified in dialogue between the principal contractor and the CDM co-ordinator. An early identification is recommended.	**Inputs to the health and safety file** *22(1)(j)*	Not only should the input be identified but also the timing of that input. The principal contractor must ensure that inputs are provided promptly.
		The principal contractor needs to be aware that the legal perspective identifies the construction site as an inducement to children to trespass.	**Prevention of access** *22(1)(l)*	All reasonable steps must be taken to secure the site to unauthorised entry. In some situations a more robust approach is required. Where this is the case information should have been provided in the pre-construction information.
		Every site worker should be the recipient of such information.	**Provision of site induction** *22(2)(a)*	This remains the primary means of conveying essential information about the site in terms of restraints, management controls and residual hazards. As with risk assessments generally, further induction of workers is required in conjunction with the need to impart additional information associated with control change.

Continued

Initial design phase	Significant detailed design phase	Pre-construction phase	Construction phase	Post-construction phase
		This would be embraced by the site induction above.	**Provision of information and training** *22(2)(b)*	Reference is made to Regulation 13(4), which addresses: • information on risks to health and safety • measures in consequence of risk • site rules • emergency procedures • identification of competent person.
			Provision of any further information and training *22(2)(c)*	This ensures that the specifics of particular work are covered by such information and training.
		This promotes the ownership of health, safety and welfare by all who can influence it and promotes the concept of management by consensus not imposition.	**Make and maintain arrangements for co-operation with workers** *24(a)*	This furthers the concept of ownership of health and safety by all through engagement in promoting and developing health, safety and welfare measures and ensuring their effectiveness throughout the project.
			Consult with workers or their representatives *24(b)*	This consultation is directly with workers or their representatives who may not have been consulted on such matters by their employer. Stress is placed on such consultation being undertaken in good time.
			Ensure workers or their representatives can inspect and take copies of information *24(c)*	This inspection and copying relates to information the principal contractor has or which he/she should have under these regulations and is essential information for planning and managing the relevant aspects of the project. Such information is in respect of matters that affect their health, safety or welfare at the site. Exceptions to the provision of such information include: • interests of national security • such disclosure leading to contravening a prohibition • disclosure specifically relating to an individual without consent • disclosure for reasons other than health, safety and welfare, which could cause substantial injury to his undertaking, etc. • disclosure obtained for bringing, prosecuting or defending legal proceedings.

Initial design phase	Significant detailed design phase	Pre-construction phase	Construction phase	Post-construction phase

8.2
PRINCIPAL CONTRACTOR CHECKLIST

PRINCIPAL CONTRACTOR CHECKLIST *(Sheet 1 of 6)*

Contract

No.	Reg.	Stage	Procedure	Description	Timing of action	Action req. Yes No	Initials	Date actioned
A	4(1)(a)	Pre-tender	**APPOINT COMPETENT DUTY HOLDERS**	All duty holders, i.e. designer(s) and contractor(s) appointed directly by the principal contractor, must be assessed as competent.	Before appointment and often well in advance via the selection procedures associated with the Approved List.			
				The principal contractor must be satisfied that the person/organisation appointed has the necessary competence to perform the function and that all individuals are either competent or under the supervision of a competent person.				
				Reliance cannot solely be placed on the self-certification process.				
				The rigour of the assessment must relate to the complexity of the project.				
B	4(1)(b)		**SELF-CERTIFICATION**	In accepting such an appointment the duty holder self-certifies their competence.				
				This applies to the appointment of the principal contractor himself by the client.				
C	20(2)(b)	Before the relevant stage of construction	**RECEIVE PRE-CONSTRUCTION INFORMATION**	Pre-construction information is to be issued promptly by the CDM co-ordinator to the principal contractor.	To enable the principal contractor to effectively embrace relevant information for the development of his safe and suitable systems of work as articulated in the construction phase plan. This should have been received at the tender stage to allow each tendering contractor to account for the health and safety management resources required and to reflect that provision in the tender price submitted.			
					Once appointed the principal contractor has to develop controls and systems in respect of such information.			

Continued

No.	Reg.	Stage	Procedure	Description	Timing of action	Action req. Yes	Action req. No	Initials	Date actioned
D	23(1)(a)	Pre-construction	**PREPARE A CONSTRUCTION PHASE PLAN**	The principal contractor is under an obligation to prepare a construction phase plan sufficient to ensure that construction is planned, managed and monitored as far as is reasonably practicable such that construction can start without risk to health or safety.	No construction work starts (including preparatory work, site clearance and investigations) until the construction phase plan has been deemed sufficient by the client (Regulation 16).				
E	22(1)(k)	Construction phase	**DISPLAY NOTIFICATION TO THE HSE (OR THE OFFICE OF RAIL REGULATION)**	Notification (commonly referred to as the F10) to the HSE or Office of Rail Regulation must be displayed in a readable condition in a position where it can be read by any worker engaged in the construction work. Usually displayed in the site office.	Throughout the construction phase.				
F	23(1)(b)	Construction phase	**REVIEW, REVISE AND REFINE THE CONSTRUCTION PHASE PLAN**	This dynamic document must continue to be developed in line with the programme of works. Arrangements must be constantly monitored to ensure effectiveness and extended to embrace the upcoming activities associated with the site momentum.	Throughout the construction phase.				
G	23(1)(c)	Construction phase	**IMPLEMENT THE CONSTRUCTION PHASE PLAN**	The construction phase plan needs to embrace all persons who may be affected by the work. It therefore has wider implications on an occupied site. Failure to identify ensures lack of control.	Throughout the construction phase.				
H	23(2)	Construction phase	**TAKE REASONABLE STEPS TO ENSURE THE CONSTRUCTION PHASE PLAN IDENTIFIES RISKS AND CONTROL MEASURES IN RESPECT OF HEALTH AD SAFETY**	Constant vigilance is required and a proactive dynamic response demanded. Similar duties apply to snagging work after project handover.	Throughout the construction phase.				

Continued

PRINCIPAL CONTRACTOR CHECKLIST (Sheet 2 of 6)

Contract

No.	Reg.	Stage	Procedure	Description	Timing of action	Action req. Yes	No	Initials	Date actioned
I	5(1)(a) and 5(1)(b)	All stages	**CO-OPERATE AND SEEK CO-OPERATION**	All duty holders must co-operate with all persons involved in construction work at the same or an adjoining site.	Particularly at the start of the project but actively throughout the project period.				
				Such co-operation must be actively sought.	Throughout the construction period.				
J	5(2)		**REPORTING**	All duty holders must report anything likely to endanger themselves and/or others to those in control. **The principal contractor must maintain open lines of communication**	Throughout the construction period.				
				Note: Your project is itself an adjoining site.					
K		All stages	**IDENTIFICATION**	There is a continuing need to identify all those with whom the duty holder needs to co-operate.	Particularly at the start of the project but actively throughout the project period.				
				This could well relate to overlapping projects, etc.					
L	6	All stages	**CO-ORDINATION**	Duty holders must co-ordinate their activities with other duty holders to ensure the health and safety of those carrying out and/or affected by such construction work.	Particularly at the start of the project but actively throughout the project period.				
				The principal contractor is the key co-ordinator in respect of site activities.					
				Additional vigilance is required where construction work is done on occupied sites.					
M		All stages	**IDENTIFICATION**	There is a continuing need to identify all those with whom the principal contractor needs to co-ordinate activities.	Particularly at the start of the construction period but actively throughout the project period.				
				This could well relate to overlapping projects, etc.					
				Co-ordination, like co-operation and communication, is an essential component of effective project management.					

Continued

No.	Reg.	Stage	Procedure	Description	Timing of action	Action req. Yes	Action req. No	Initials	Date actioned
N	7	All stages	**ACCOUNT FOR GENERAL PRINCIPLES OF PREVENTION**	The basic tenets of health and safety management must be embraced by all duty holders and applied to all work activities on site. Such compliance also links with the management of construction duties, particularly under Regulation 13(4)(c), and reinforces the requirements of the MHSW Regulations 1999.	Throughout the project period.				
O	22(1)(a)	Construction phase	**PLAN, MANAGE AND MONITOR**	The construction phase must be managed at all times accounting for the general principles of prevention and embracing the need for co-operation and co-ordination between persons concerned in the project or affected by it.	Throughout the construction phase.				
P	22(1)(b)	Construction phase	**LIAISE WITH CDM CO-ORDINATOR**	Liaison is required with the CDM co-ordinator in matters relating to design or design change. This often requires change to existing control arrangements and can be aggravated through the confines of the construction programme itself.	Throughout the construction phase.				
Q	22(1)(c)	Construction phase	**ENSURE WELFARE FACILITIES ARE SUFFICIENT**	Welfare facilities must comply with the requirements of Schedule 2. They must be regularly serviced and continue to function effectively. The frequency of servicing should be identified in the construction phase plan.	From the first day of construction and throughout the construction phase.				
R	22(1)(d)	Construction phase	**DRAW UP SITE RULES**	Site rules are an implicit part of site management and controls and need to be conveyed to all workforce members via the induction programme. Such rules must be site specific.	Such rules and the induction programme itself need to be re-addressed in line with the change in work procedures and re-conveyed to workforce members as appropriate. Rules and related information needs to be in a comprehensible language and thus reflect the ethnic mix on site.				

Continued

PRINCIPAL CONTRACTOR CHECKLIST (Sheet 4 of 6)

Contract						Action req.			
No.	Reg.	Stage	Procedure	Description	Timing of action	Yes	No	Initials	Date actioned
S	22(1)(e)	Construction phase	**GIVE REASONABLE DIRECTIONS**	The principal contractor is the directing influence on site and reasonable directions enable him to discharge his duties Such directions extend to all those on the construction site throughout the period	Throughout the construction phase				
T	22(1)(f)	Construction phase	**INFORM EVERY CONTRACTOR OF THE MINIMUM PERIOD OF MOBILISATION**	This is the minimum period for planning and managing before the contractor begins construction work. This period should be ratified via dialogue between the principal contractor and the contractor.	In advance of preparation for that particular element of work.				
U	22(1)(g)	Construction phase	**CONSULT CONTRACTORS BEFORE FINALISTION OF THE CONSTRUCTION PHASE PLAN**	This facilitates the ownership of health and safety by those parties who can influence it and serves to promote the team-based approach and further builds on the expertise of the party concerned.	In advance of preparation for that particular element of work.				
V	22(1)(h)	Construction phase	**ALLOW ACCESS TO RELEVANT PARTS OF CONSTRUCTION PHASE PLAN**	This is transparency of management and allows the contractor access to the key site control document so that he can prepare properly for the related work activity.	Before the commencement of related construction work.				
W	22(1)(i)	Construction phase	**PROVIDE FURTHER INFORMATION**	Such information relates to compliance with welfare provision (Schedule 2) and information to enable work to be performed by him without health and safety risk to any person.	Throughout the construction phase.				
X	22(1)(j)	Construction phase	**IDENTIFY INPUTS TO THE HEALTH AND SAFETY FILE**	The identification of contractor input to the health and safety file should be done after discussion between the CDM co-ordinator and the principal contractor.	Such inputs need to be identified as early as possible, including the timing of the input.				

Continued

Contract						Action req.			
No.	Reg.	Stage	Procedure	Description	Timing of action	Yes	No	Initials	Date actioned
Y	22(1)(l)	Construction phase	PREVENT ACCESS BY UNAUTHORISED PERSONS	Within the principal contractor's remit of managing the site comes the need to ensure the site is secure from trespassers. Reasonable steps must be taken and relevant information should be provided in the pre-construction information tender stage document to advise if a more robust approach is required because of environmental challenges.	Throughout the construction phase regardless of the frequency of vandalism.				
Z	22(2)(a)	Construction phase	PROVIDE SITE INDUCTION	Every worker must be the recipient of not only site induction information but also information and training needed to undertake work in a safe and healthy manner.	Throughout the construction phase and repeated if circumstances change.				
AA	22(2)(b)	Construction phase	PROVIDE INFORMATION AND TRAINING	This should include: • information on risks to health and safety • measures in respect of risk control • site rules • emergency procedures • identification of competent person as referred to in Regulation 13(4).	Throughout the construction phase.				
AB	22(2)(c)	Construction phase	PROVIDE FURTHER INFORMATION AND TRAINING	The principal contractor shall provide every worker with any further information and training specifically required for particular work so that it can be carried out safely.	Throughout the construction phase.				
AC	24(a)	Construction phase	MAKE AND MAINTAIN ARRANGEMENTS FOR CO-OPERATION WITH WORKERS	This is the promotion of health and safety responsibility through ownership and consolidates the approach outlined in the Health and Safety at Work, etc. Act 1974.	Throughout the construction phase.				
AD	24(b)	Construction phase	CONSULT WITH WORKERS OR THEIR REPRESENTATIVES	This consultation is directly with workers or their representatives who may not have been consulted on such matters by their employer.	Throughout the construction phase.				
AE	24(c)	Construction phase	ENSURE WORKERS OR THEIR REPRESENTATIVES CAN INSPECT AND TAKE COPIES OF INFORMATION	This inspection and copying relates to information the principal contractor has or which he should have under these regulations and is considered essential information for planning and managing the relevant aspects of the project in respect of workers' health, safety or welfare at the site. Exceptions occur (see p. 121).	Throughout the construction phase.				

PRINCIPAL CONTRACTOR CHECKLIST (Sheet 6 of 6)

Section 9
DOCUMENTATION

9.1 Introduction

In addition to the notification to the enforcing authority (F10), the documentation directly identified within the CDM Regulations 2007 and the corresponding ACoP consists of the information process identified as pre-construction information together with the construction phase plan and the health and safety file.

In association with industry practice the above are also supplemented by project risk (health and safety) registers, design risk and work place risk assessments and method statements.

The main purpose of documentation is to facilitate communication at all relevant stages of the process but it also provides input to the trail of accountability as evidence of the discharge of duties. It is imperative that the duty holder appreciates the true purpose of documentation and continually strives to 'manage the risk and not the paperwork'.

For all purposes documentation should be focused and succinct, with a site-specific approach related to significant and principal issues. This latter focus strays away from the ACoP thrust but in so doing offers the opportunity to embrace the strategic interventionist policies of the enforcing authorities. It also acknowledges the subjectivity synonymous with the word 'significant' and therefore allows the capture of current concerns. However, the duty holder must at all times exercise due care and diligence in ensuring irrelevant content is filtered out.

There is always the danger that key messages can be lost through excess and superfluity.

Additionally, there is an inter-relationship between all the documentation and it should be viewed collectively as a portfolio of team contributions for the enhancement of health and safety management throughout the project.

Prior to the advent of the CDM Regulations 2007, a major criticism was the excessive paperwork linked to associated documentation and the acknowledgement from all that such an approach did not improve the health and safety risk management process. This situation must be avoided so that management systems can support the risk management process without enslaving the duty holder to bureaucratic excesses.

The criticism of the excessive paperwork could be seen by the cynic as arising from an abdication of managerial responsibility. Hence all duty holders, including the enforcing authorities, have roles to play in improving the process. As noted by Stuart Nattrass, the Chief Inspector of Construction, at the launch of the CDM Regulations 1994, the usefulness of paperwork is in inverse proportion to the amount.

For notifiable projects it is the CDM co-ordinator who is charged with the overall management of the pre-construction information and whilst conventional systems would be adequate for smaller projects, recourse to more sophisticated computerised systems must be considered for more complex project arrangements. On non-notifiable projects document management falls to the client, who inevitably requires help from his professional team.

9.2 Pre-construction information

This must be seen as an information process that gathers momentum throughout the project and should therefore be viewed as an information highway that is not time barred. Unlike the earlier regulations **pre-construction information is required at all stages of all projects, not only at the tender stage**. It should not preclude additional information going out to relevant duty holders in line with the iterative nature of project development.

Such information has to go out to all designers and contractors who have been or who might be appointed by the client, with some information being provided at the construction stage reflecting procurement strategy and project development.

As emphasised in the ACoP:

'The level of detail in the information should be proportionate to the risks involved in the project.'

Those responsible for providing the information must decide on its relevant content.

Such a process must deliver relevant information on both non-notifiable and notifiable projects and subsequently imposes duties on clients (Regulation 10(1)) and CDM co-ordinators (Regulation 20(2)(b)). It should also be noted that all other duty holders should contribute to the content of such information.

Appendix 2 of the ACoP offers a format for the pre-construction information with an introductory paragraph that does not distinguish between notifiable and non-notifiable projects.

Hence, whilst the format bears a distinct resemblance to the obsolete pre-tender health and safety plan (CDM Regulations 1994) the inference by way of the presentation of Appendix 2 is that relevant sections could also be used for all stages of information flow. A similar selection process will accompany non-notifiable projects.

For notifiable projects the CDM co-ordination team will manage this process, whilst on non-notifiable projects the extent of guidance from the assembled professionals will depend on the construction experience of the client.

Undoubtedly, CDM co-ordinators associated with notifiable projects could offer invaluable support to the lay client on non-notifiable projects under the guise of a CDM advisor.

As shown in Figure 9.1, because of the information flow there is a need to manage and qualify each stage release. Inevitably not all recipients will get the same information, which is decided by the nature of the project.

As well as the process diagram, an example of a pre-construction information-tender stage document is provided for consideration. The format of this document complies with that set out in Appendix 2 of the ACoP (L144).

Figure 9.1

FLOW OF
PRE-CONSTRUCTION
INFORMATION

PRE-CONSTRUCTION INFORMATION: TENDER STAGE

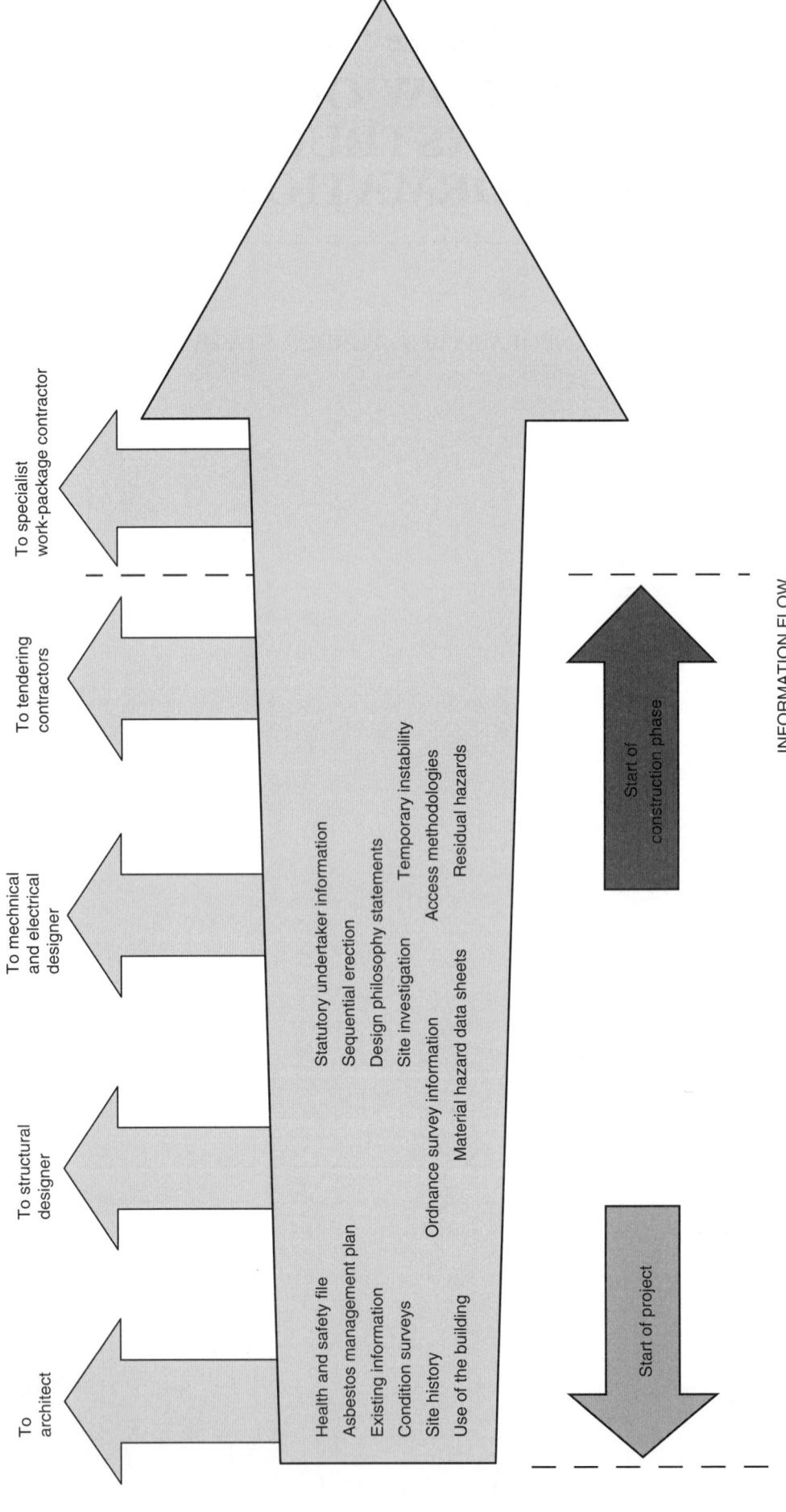

Figure 9.1 Pre-construction information

Pre-Construction Information–Sampe Document

Pre-Construction Information: Tender Stage
for
Development Project

Business Lane
Commercial Park
New Town
FC12 54Q

Status	Revision	Date
Tender	0	Today's date *(useful to facilitate document management)*

Contents

1. Description of project

Project description and programme details	
Description	The project consists of the demolition of existing single-storey buildings (four) and the construction of a five-storey conference and communication centre, together with associated infrastructure works.
Key dates	Planned start: 3 September 2007 Completion: 30 May 2008 Sectional completion: 29 February 2008
Mobilisation time	Four weeks from appointment of principal contractor
Duty holders	*Client* ADG Holdings Ltd Clearview Lane Holden NM76 9BC Contact: Alan Bold Tel.: e-mail: *Designer: Architect* GH Associates LLP West of Everywhere Essex EX8 7JI Contact: Bill Spret Tel.: e-mail: *Designer: Structural* Analysts 'R Us Ruby Way Tabsworth TX21 4GF Contact: Andy Lowe Tel.: e-mail: *CDM co-ordinator* SCMC Ltd 9 Aspect Place Green Acres New Town FF72 8AA Contact: Anton Dec Tel.: e-mail: *Other consultants* To be appointed
Use as a workplace	The project consists of five-storey office accommodation, which will house 750 staff and will need to comply with the Workplace (Health, Safety and Welfare) Regulations 1992.
Existing records	Service drawing: QU/services/1329/Rev 5 (October 2005) Geotechnical report: Reference no. 12/PS/Oct/2006/Rev 2 Asbestos Management Plan: ASBMP/1234/Sept. 2006

2. Client's considerations and management requirements

(a) Arrangements for:

(i) Planning and managing construction work, including any health and safety goals for the project

- The principal contractor must establish a benchmark standard for the monitoring of health and safety management on this project, e.g.:
  – all-incident frequency rate of 4.0/100,000 person hours
  – reportable accident frequency rate of nil
- All contractors on site will be expected to achieve a similar minimum standard. Site audits and inspections are to feed into progress meeting reports.

The construction phase plan (formerly the construction phase health and safety plan) developed from the pre-construction (tender stage) information must be submitted to the CDM co-ordinator not less than one week before the proposed start date for construction work.

No construction work is to commence until confirmation has been received in writing from the client that the construction phase plan is sufficiently developed in compliance with Regulations 23(1)(a), 23(2) and 22(1)(c).

The sufficiency of the construction phase plan in respect of the above is dependent on the inclusion of suitable method statements for:

- site security
- traffic management arrangements within an operational facility
- welfare arrangements
- asbestos removal (roof sheeting in obsolete buildings)
- demolition
- co-operation with overlapping projects (refer to (ii) below)
- fire action plan
- site waste management plan
- noise (75 dB limit)
- structural stability for steelwork erection.

(ii) Communication and liaison between client and others

The client's representative is:
  Alan Bold
  Tel.:
  e-mail:

Industrial park management is under the jurisdiction of:
  John Brown/Estates Facilities Manager
  Tel.:
  e-mail:

New car park project (January 2008 to April 2008):
  Principal contractor: Mactar Group
  Contact: Kilroy Smith
  Tel.:
  e-mail:

Continued

    (iii) Security of the site
- Industrial park site security is under the management of the Secure With Us Group based at the main access gate (tel.:   ).
- All delivery vehicles must be prominently marked.
- All 'white vans' will be searched.
- All vehicles and individuals must sign in each time of entry to the industrial park itself.

    (iv) Welfare provision
- Both foul and storm water drains are available to connect into adjacent to the north-east boundary (the allocated location for site accommodation).
- No electrical supply is provided.
- Water supply connection is available at the entrance to the construction site. Contact John Brown/Estates Facilities Manager (tel.: 01234 558877). 48-hour notice required for connection.

(b) Requirements relating to the health and safety of the client's employees or customers or those involved in the project

    (i) Site hoarding requirements
- Any site hoarding arrangements must be colour co-ordinated to match the two-tone colour scheme represented on the client's corporate logo.
- Vision panels must be incorporated (two per side) into all hoardings.
- Access to the stand-by transformer station on the western flank of the site (ref. site drawing: GA/Estates/12324/May 2007) must be maintained at all times for facilities management staff.

    (ii) Site transport arrangements or vehicle movement restrictions
- All vehicles and foot traffic must enter via central security control, situated at the Central Avenue/B5678 intersection.
- Twelve-tonne weight limit on access road B.
- One-way traffic schemes exist throughout the industrial park.
- Speed restrictions of 10 mph exist throughout the main site area.

    (iii) Client permit to work systems
- An Authorisation to Work permit is required at the start of every week. This is to be issued by the facilities management team. Contact John Brown/Estates Facilities Manager (tel.: 01234 558877).
- Such permits are required for all service connections to existing mains.

    (iv) Fire precautions
- A fire action plan co-ordinated with the emergency evacuation procedures operated by the facilities management team must be incorporated into the construction phase plan. Contact John Brown/ Estates Facilities Manager (tel.: 01234 558877).

Continued

(v)   Emergency procedures and means of escape
- Refer to (iv) above.
- The nearest A&E hospital is located five miles away:
  General Infirmary
  Bloodlet Lane
  New Town
  FC12 54Q
  Tel.: 01987 345 765
- A route map from the site to the hospital is to be included with the construction phase plan.

(vi)   'No-go' areas or other authorisation requirements for those involved in the project
- All service connections require a Disruption to Service Provision order issued by the facilities management team. Contact Bill Rew (mobile:). Seven days' advance notice is required.
- No contractor is to enter the sterile laboratory areas identified as buildings LAB 1, LAB 2, LAB 3 and LAB 4.
- Such entry will result in instant dismissal from the industrial park.

(vii) Any areas the client has designated as confined spaces
- Sump pit in pumping station.
- Manholes FS 23, FS 27 and FS 30 (refer to Service Drawing QU/services/1329/ Rev 5 (October 2005).

(viii) Smoking and parking restrictions
- No parking allowed on any service road.
- The industrial campus is a no-smoking area other than in designated rooms annexed to restaurant/cafeteria areas.

3. Environmental restrictions and existing on-site risks

(a) Safety hazards, including:

(i) Boundaries and access
- Access to the site is one mile from Junction 27 of the M28, off the A37 Crockett to Threshold road, and then via the B5678.
- The B5678 is a single-lane road with passing places.

(ii) Restrictions on deliveries, waste collection or storage
- Material deliveries must take place outside the high density traffic periods of 0730 hours to 0900 hours and 1600 hours to 1800 hours.

(iii) Adjacent land uses
- To the north-east of the site at the A37/B5678 junction is situated the new Manor Way Comprehensive School (1200 pupils).
- A Site Waste Management Plan is to be provided as part of the construction phaseplan (for projects over £250 000).

(iv) Existing storage of hazardous materials
- None

(v) Location of existing services
- As identified on Service Drawing QU/services/1329/Rev 5 (October 2005).

(vi) Ground conditions
- Geotechnical report reference no. 12/PS/Oct/2006/Rev 2 has identified the presence of 'running sand', encountered in boreholes 2 and 14.

(vii) Information about existing structures
- Asbestos roof sheeting has been identified in the obsolete buildings.
- Dry rot infestation affects the roof members of these buildings.

(viii) Previous structural modifications
- Not applicable.

(ix) Fire damage, etc.
- Building B23 has been extensively damaged by fire, which is known to have affected the main roof support members.

(x) Difficulties in relation to existing plant and equipment
- The access route to the foul pumping station has a height restriction limitation of 3.50 metres clearance and a width restriction of 3.25 metres.

(xi) Health and safety information contained in earlier design, construction or as-built drawings
- Nothing applicable.

(b) Health hazards, including:

(i) Asbestos
- Roof sheeting in obsolete buildings.

(ii) Existing storage of hazardous materials
- Not applicable.

Continued

(iii) Contaminated land, including results of surveys
- Geotechnical report reference no. 12/PS/Oct/2006/Rev 2 has identified the presence of:
  – trace of phenols in trial pit TP3
  – hydrocarbon deposits in borehole 27.

(iv) Existing structures containing hazardous materials
- Sharps and needles in associated ground floor areas.

(v) Health risks arising from client's activities
- Traffic movement arising out of operational use.

4. Significant design and construction hazards

(a) Significant design assumptions and suggested work methods, sequences or other control measures
- Sequential erection details for temporary stability (refer to Appendix 3).
- Hot work minimised via full length stanchions and bolted connections.
- All wet cement products are based on low chromium cement specifications.

(b) Arrangements for co-ordination of ongoing design work and handling design changes
- All Architect's Instructions/Variation Orders having a design implication must be forwarded to the CDM co-ordinator to assess the impact on the development of the construction phase plan. *(Regulation 20(1)(c)(ii) requires the CDM co-ordinator to liaise with the principal contractor regarding 'any design development which may affect planning and management of the construction work'.)*
- Co-operation/co-ordination must be established between permanent and temporary design processes.

(c) Significant and principal risks identified during design
- Site security.
- Traffic management arrangements within an operational facility.
- Welfare arrangements.
- Asbestos removal (roof sheeting obsolete buildings).
- Demolition.
- Co-operation with overlapping projects (see later).
- Fire action plan.
- Site waste management plan.
- Noise (75 dB limit).
- Structural stability for steelwork erection.
- Work at height: atrium roof, roof-top parapets, plant rooms, cladding.

(d) Materials requiring particular precautions
- Epoxy resin associated with holding down bolts.
- Manual handling:
  - kerbs: weights are typically 32 and 65 kg for 'specials'
  - piling shells: weights estimated to be 35 kg.

5. The health and safety file

It is a requirement of the regulations that the principal contractor, in discussion with the CDM co-ordinator, identifies the input required of contractors for inclusion in the health and safety file, and implements an effective management system by which such information is promptly provided to the CDM co-ordinator *(refer to principal contractor duty in Regulation 22(1)(j))*.

The client requires one copy of the health and safety file in hard copy format together with a CD rom back-up with drawings in pdf format.

Format and content *(refer to clause 263 of the ACoP (L144)*:

    (a) A brief description of the work carried out.

    (b) Residual hazards and how they have been dealt with (for example surveys or other information concerning asbestos, contaminated land, water-bearing strata, buried services, etc.).

    (c) Key structural principle (e.g. bracing, sources of substantial stored energy, including pre- or post-tensioned members) and safe working loads for floors and roofs, particularly where these may preclude placing scaffolding or heavy machinery there.

    (d) Hazardous materials used (for example lead paint, pesticides, special coatings which should not be burned off, etc.).

    (e) Information regarding the removal or dismantling of installed plant and equipment (for example any special arrangements for lifting, order or other special instructions for dismantling, etc.).

    (f) Health and safety information about equipment provided for cleaning or maintaining the structure.

    (g) The nature, location and markings of significant services, including underground cables, gas supply equipment, fire-fighting services, etc.

    (h) Information and as-built drawings of the structure, its plant and equipment (e.g. the means of safe access to and from service voids, fire doors and compartmentalisation, etc.).

Signed: Dated:

6. Appendices (may include the following, which are not displayed in full here)

1. Site boundary plan

2. Project risk register

3. Sequential erection procedure

9.3   Construction phase plan

This is the control document for any notifiable project and is owned by the principal contractor who is in control of the corresponding construction site. It should articulate all the on-site management arrangements and controls for providing safe and suitable systems of work at all stages of construction. It is a continually developing document compatible with the programme of works and initially builds on the information provided within the corresponding pre-construction information tender stage document. As such it is housed on site, with the principal contractor allowing relevant access to it for his contractors and sub-contractors (Regulation 22(1)(h)).

However, the client, with the facilitation provided by the CDM co-ordinator, must ensure that the plan has been sufficiently developed to allow the construction phase to start (Regulation 16). This is a critical gateway of client control and conditional acceptance of such a document is unacceptable.

No aspect of the construction phase work on site should start until the construction phase plan has been deemed sufficient by the client, i.e. sufficient to ensure that the early stages of construction are adequately controlled, with strong emphasis being placed on adequate welfare facilities on site from day one.

At this juncture the client, CDM co-ordinator and principal contractor need to be clear that site investigation works, preparatory works and site clearance are clear indicators that the construction phase has started and such aspects of the project must all be fully embraced within the project itself and thus compliant with the controls to be exercised by both the client and the principal contractor.

Appendix 3 of the ACoP provides the format for such a document with sub-sections being used relevant to the project.

Whilst such a document is not required on non-notifiable sites, workplace management controls must still be in place and statutory duties to the Management of Health and Safety at Work Regulations 1999 and other construction-related workplace legislation still require a compatible response from duty holders.

9.4   Health and safety file

This document accompanies the notifiable project and is handed over to the client at the end of the construction phase by the CDM co-ordinator, who is now responsible for its preparation.

This does not necessarily translate into the CDM co-ordination team actually putting the document together but they undoubtedly are directly responsible for its preparation, whatever agreements are developed between duty holders to ensure that this is done. Failure to produce the health and safety file at construction completion constitutes non-compliance with the regulations, with direct implications for the CDM co-ordination team. The management process associated with the health and safety file commences immediately on the appointment of the CDM co-ordinator and a proactive approach needs to be adopted continually throughout all stages of the project.

There is no format outlined in appendices to the ACoP but paragraph 263 provides guidance on content.

Whilst a health and safety file accompanies a notifiable project, clients need to be aware that work on non-notifiable projects could require an amendment/modification to an existing health and safety file. Management arrangements must be capable of a suitable response to such situations.

9.5 Project risk register

Project risk registers have been associated with project management control for a considerable time, mainly directed at financial risk and exposure.

As a recognised methodology they offer a useful means of extension into health and safety risks, as mentioned in Chapter 5. Indeed health and safety risk exposure could well represent the major risk issue on projects and could be quantified in financial terms if so required.

In the above project management context such an approach should embrace all duty holders and offers a means of integrated team promotion through empowerment and contribution to the ownership of health and safety management issues.

The one dormant question relates to the ultimate ownership of this document. This could fall within the remit of the client, project manager, CDM co-ordinator, lead designer and/or principal contractor, depending on project arrangements, but needs to be resolved at the start. All duty holders, however, contribute at the requisite stages and thus such a facility promotes integration through dynamic input, beginning with the client.

The further benefit is that the project risk register accompanies the project and invites duty holder contribution in an iterative sense, unlike the limitations relating to the design risk assessment, which focuses only on one duty holder facet. Such an approach also acknowledges the nature of project development and once visibility is given to the identified risk, associated ownership ensures that it remains 'open' on the register until it is closed out via a satisfactory response. This response may only occur some time later after aspects of constructability/maintainability have been fully resolved, often with a principal contractor/contractor input.

This itself provides further endorsement for the early appointment of the principal contractor, although such an appointment would depend on the procurement strategy.

Management of this process must ensure that all significant/principal issues remain on the register to provide continuing visibility of relevant parameters and allow full appreciation of the project development trail.

Additional contributions to the project risk register are encouraged simultaneously with duty holder contributions throughout the project and thus a dynamic approach could be fostered via an intranet link. This counters the criticism of the belated response and should provide a real-time scenario for more effective management control.

An example of a project risk register follows.

9.1
PROJECT RISK REGISTER (HEALTH AND SAFETY)

PROJECT RISK REGISTER (HEALTH AND SAFETY)

Example				Sheet 1 of	Date:	Revision: 1	Author:	
Ref. no.	Duty holder	Risk description	Assessment	Mitigation/controls	Assessment	Ownership	Future action	Status
01	Client	**Asbestos (obsolete buildings)**	**High**	Asbestos management plan: Remove	Low	Demolition contractor	Removed	Completed
	Client	**Unknown services, etc.**	**High**	Service information	Low	Client	Provided	Completed
01	Client	**Demolition**	**High**	Existing drawings	Medium	Principal contractor	Methodology Written method statement Construction phase plan	Live/ongoing
		Flooding	**Medium**	Gauging station information	Medium	Principal contractor	Methodology Construction phase plan	Live/ongoing
02	Designer: Structural	**Structural stability**	**Medium**	Sequential erection	Low/medium	Principal contractor	Methodology Construction phase plan Health and safety file Specification	Live/ongoing Strategies being developed
		Hand/arm vibration syndrome	**Medium**	Pile preparation by chemical expansion	Low	Designer/principal contractor	Methodology	Live ongoing
		Work at height	**High**	Pre-fabrication Ground floor assembly	Low	Designer/principal contractor	Shop details Methodology	Live/ongoing
		Work at height	**High**	Permanent formwork	Low	Principal contractor	Methodology Construction phase plan	Live/ongoing
03	Designer: Mechanical and engineering	**Structural integrity of roof (air conditioning units)**	**Medium**	Structural survey	To be assessed	Designer: Mechanical and engineering	Awaiting report	Live/ongoing
04	Principal Contractor	**Noise limitations**	**Medium**	Plant choice Phasing of works	Medium	Piling contractor	Methodology CP Plan	Live/ongoing
05	Contractor	**Manual handling**	**Medium**	Mechanisation Automation	Low	Contractor	Methodology Construction phase plan Health and safety file	Live/ongoing

Section 10
CHECKLISTS AND AGENDAS

The checklist concept is often viewed as contrary to the aims of lateral thinking and can introduce constraints to lateral thinking and creativity, both essentials for health and safety management control.

In this sense checklists and agendas should not be seen as an end in themselves but as a useful aide-memoire to be referred to at an opportune time. They should also not be considered as exhaustive or necessarily comprehensive for review, and monitoring is essential to ensure they cover the requisite points and evolve compatibly with industry best practice.

Within any competent team rests a wealth of experience, which provides a useful reservoir of focus and appreciation, for input to the above. This is the foundation of contribution in the positive sense but in the negative sense can also introduce 'baggage' through complacency and prejudice. This needs to be avoided and challenged at all stages.

Across the continuum of the construction field many areas have specific and particular needs that need to be addressed situation by situation. Checklists and agendas need to be tailored and/or modified accordingly.

The following checklists are offered not as absolutes but as reference points for further development. The relevant duty holder is identified by **x**, whilst **(x)** signifies the necessary involvement of other duty holder groups. Many of the duties to be exercised recur throughout and must be revisited at numerous stages, e.g.:

- duty holder identification: continuing competence, co-operation with those on the site and those on adjoining sites, co-ordination of activities with others, general principles of prevention, maintaining and review of management arrangements
- pre-construction information: avoidance of foreseeable risks, sufficiency of information, design undertaken outside Great Britain, plan, manage and monitor, suitable and sufficient advice, suitable and sufficient arrangements, management of the health and safety file, liaison.

10.1
CHECKLIST FOR NON-NOTIFIABLE PROJECTS

CHECKLIST FOR NON-NOTIFIABLE PROJECTS (*Sheet 1 of 6*)

Reg.	Action	Duty holders			Comment
Pre-start		Client	Designer	Contractor	
2	**Is the work construction work?**	x	(x)		**There are few exceptions, but distinguish between:** • *'survey'* and *'investigation'* work • *'exploration for or extraction of mineral resources or activities preparatory thereto carried out at a place where such exploration or extraction is carried out'* • additional exemptions outlined in paragraph 13 of the ACoP. This is the earliest question to be asked by the client before a commitment is made to comply with the CDM Regulations 2007. For the lay client the first point of contact is usually the designer (**x**), who will need to verify the nature of the work.
4(1)(a)	**Is the team competent?**	x	x	x	All duty holders appointed or engaged must be assessed for competence.
	Have you . . .	Client	Designer	Contractor	
	Identified duty holders?	x	x	x	Awareness must be exercised throughout the project since the duty holder list is a project dynamic and will change as duty holders are appointed/engaged. It should be noted that all sub-designers are designers and all sub-contractors are deemed to be contractors for the purpose of the CDM Regulations 2007.
4(1)(b)	**The competence to fulfil your duties under any of the relevant statutory provisions?**	x	x	x	In accepting an appointment or engagement the duty holder self-certifies their competence.
4(1)(c)	**Established that all workers carrying out or managing design or construction work are competent or are under the supervision of a competent person?**	x	x	x	Organisations must structure teams so that supervision is provided by a competent person. Such a structure would need to reflect the complexity of the project.
(5)(1)(a)	**Sought the co-operation of any other relevant person involved in construction on this or an adjoining site in order to fulfil your duties?**	x	x	x	Such co-operation is essential for effective health and safety management. This is also a project dynamic and the relevant contact list will change throughout the project. **Note:** • Your site is itself an adjoining site. • This applies to any other person concerned in any project involving construction work not just duty holders.
	Identified such parties with whom you need to co-operate?	x	x	x	For example overlapping projects, shared access, operational management sites, partial handover, sectional completion, etc.

Continued

Reg.	Action	Duty holders			Comment
Pre-start		**Client**	**Designer**	**Contractor**	
5(1)(b)	**Co-operated with other relevant persons concerned in construction work at this or an adjoining site?**	x	x	x	Effective co-operation is facilitated by the identification of roles and responsibilities supported by adequate lines of communication.
5(2)	**Reported anything likely to endanger the health and safety of yourself or others to those in control?**	x	x	x	Proactive monitoring and reporting remain the key to effective control systems.
6	**Co-ordinated your activities with other duty holders to ensure the health and safety of those affected by and carrying out construction work?**	x	x	x	Such co-ordination is dependent on a lead role being undertaken by the 'lead designer' and/or the 'main contractor'. The co-ordinating roles should reflect the duty holder effectively positioned in respect of the procurement strategy for the project.
7	**Accounted for the general principles of prevention by all those involved in the design, planning, preparation and construction of the project?**	x	x	x	Such principles are outlined in Appendix 7 of the ACoP All duty holders must ensure that such principles of prevention are applied throughout all phases of the project
8	**Elected one of the group of clients to represent the group of clients or is this not required on this project?**	x			Such an arrangement can facilitate communication within the project context, e.g. on a development site with numerous independent client interests.
8	**Received written confirmation of the client election from all other clients?**	x			All client parties need to be identified and confirmation received. Duties in respect of co-operation, provision of relevant pre-construction information and health and safety information remain to be discharged by all clients regardless of who is elected to represent them.
9(1)	**Taken reasonable steps to ensure that management arrangements made by duty holders are suitable to ensure:** ● **safe construction** ● **welfare facilities are complied with** ● **relevant structures are compliant with the Workplace (Health, Safety and Welfare) Regulations 1992?**	x			Such steps include confirmation that suitable management infrastructures (including allocation of sufficient time and other resources) accompany duty holder roles throughout the project. This includes the establishment of roles and responsibilities together with key lines of communication. Particular focus must be provided in pre-start meetings with all duty holders.

Continued

CHECKLIST FOR NON-NOTIFIABLE PROJECTS (Sheet 2 of 6)

Reg.	Action	Duty holders			Comment
		Client	Designer	Contractor	
Pre-start					
9(2)	**Ensured management arrangements are being maintained and reviewed throughout the project?**	x			The client must ensure that such arrangements continue to be effective and must therefore address and re-address such issues in: • pre-start meetings • review meetings • progress meetings • development meetings • co-ordination meetings, etc.
10(1)	**Provided pre-construction information to all designers, and contractors who have been or may be appointed?**	x			Such information must be provided promptly. **Note**: This is not pre-construction phase information and therefore could be provided during the construction phase, e.g. to a specialist work package designer/contractor appointed during the construction period.
10(2) and 10(3)	**Identified such pre-construction information in the client's possession?**	x	(x)	(x)	Pre-construction information includes information in the client's possession that is reasonably obtainable, including: • relevant information about/affecting the site/construction work • relevant information concerning the proposed use of the structure as a workplace • the minimum amount of time required for mobilisation • relevant information in an existing health and safety file. In many situations the client will need guidance for the identification of such information. This information is essential to enable other duty holders to fulfil their duties and adequately resource the effective health and safety management controls associated with the discharge of their duties. For notifiable projects identification and collection are undertaken by the CDM co-ordinator; on non-notifiable projects similar advice and assistance is required for the lay client but the regulations are silent on who provides such assistance. This could come from the: • assembled team • competent person appointed under Regulation 7 of the MHSW Regulations • CDM advisor.

Continued

Reg.	Action	Duty holders			Comment
Pre-start		Client	Designer	Contractor	
11(1)	**Ensured the client is aware of his duties under these regulations?**		x		All designers have to be satisfied that the client is aware of his duties before any design work starts. In a multi-disciplinary design situation it would be sufficient that other designers receive confirmation from the lead designer that this has been done.
	Received relevant pre-construction information?		x	x	Such information is to be received from the client by all designers and those contractors who are appointed or may be appointed by the client.
11(2), 11(3) and 11(4)	**Avoided within this design foreseeable risks to the health and safety of those associated with construction, who will or may:** • **carry out such work** • **are liable to be affected by such work** • **clean windows or transparent/translucent elements** • **maintain permanent fixtures/fittings** • **use the structure as a place of work?**		x		Such avoidance is qualified by the phrase 'so far as is reasonably practicable' and serves to discharge duties within the remit of the general principles of prevention as detailed in Appendix 7 of the ACoP. This regulation requires the design teams to fully embrace the health and safety implications throughout all the phases of the construction process.
11(5)	**Designed the structure for use as a workplace in compliance with the Workplace (Health, Safety and Welfare) Regulations 1992?**		x		Design duty here is focused on the design of and materials used in the structure. Some designers would be designing outside the remit of the Workplace (Health, Safety and Welfare) Regulations 1992, since this particular regulation only refers to fixed workplaces such as offices, shops, factories and schools.
11(6)	**Provided sufficient information about aspects of the design, of the structure, or its construction or maintenance to:** • **clients** • **other designers** • **contractors** **to enable them to comply with their duties under these regulations?**		x		Adequate and sufficient information for work to be undertaken safely must be provided to other duty holders at the appropriate time. This will enable strategies and safe systems of work to be developed as well as the integration of design parameters to be appreciated. Within this context such information could be provided via: • pre-construction information • notes on drawings • additional information • health and safety file.

Continued

CHECKLIST FOR NON-NOTIFIABLE PROJECTS (Sheet 5 of 6)

Reg.	Action	Duty holders			Comment
		Client	Designer	Contractor	
Pre-start					
	Established if any design is being done outside Great Britain?	x	x		Duties exist on those who commission design work to be done outside Great Britain to ensure that all design is compliant with the requirements of Regulation 11. Such a question needs to be asked at relevant stages throughout the project.
12	**Ensured that all design prepared or modified outside Great Britain not commissioned by others complies with Regulation 11?**	x	x		This duty falls on those who appoint designers from outside Great Britain. If the commissioning route cannot be established then the duty falls back onto the client.
	Established if any aspects of the design require an amendment/modification to an existing health and safety file?	x	x	x	Work on a non-notifiable project could still affect the currency of information in an existing health and safety file. Health and safety management must ensure appropriate amendments/modifications are made.
13(1)	**Ensured the client is aware of his duties?**			x	All contractors have to be satisfied that the client is aware of his duties before any construction work starts. In a multi-contractor situation it would be sufficient that other contractors receive confirmation from the main contractor that this has been done.
13(2)	**Effectively planned, managed and monitored construction work to ensure work is carried out without risks to health and safety?**			x	This applies to all construction work directly undertaken or under the control of the contractor and is qualified by the term 'as far as is reasonably practicable'. This represents an ongoing duty.
13(3)	**Informed all contractors engaged or appointed by yourself of the minimum amount of time for planning and preparation before construction work begins?**			x	This is to ensure that the mobilisation period is realistic and compatible with the needs of the corresponding construction work. Such a minimum period of mobilisation should be ratified by the contractor engaged or appointed before work begins.
13(4)	**Provided every worker carrying out construction work under your control with:** • **relevant information and training** • **suitable site induction** • **residual risk information** • **other contractor issues** • **prevention and protection measures** • **site rules** • **emergency procedures** • **the identity of the nominated person (emergency procedures)?**			x x x x x x x x	Much of this information can be imparted through a well-structured induction programme. It should be noted that no induction programme is a once and for all situation. Individuals may have to be re-inducted as situations change in line with site progress, etc.

Continued

Reg.	Action	Duty holders			Comment
		Client	Designer	Contractor	
Pre–start					
13(5)	**Provided employees with health and safety training by virtue of Regulation 13(2)(b) of the MHSW Regulations 1999?**			x	This is training required on being exposed to new or increased risks arising out of new responsibilities, the introduction of new work equipment, the introduction of new technology and/or introduction of new systems of work.
13(6)	**Taken all reasonable steps to prevent access by unauthorised persons to the site?**			x	Construction sites remain dangerous and reasonable efforts must be made to secure them against trespassers and vandalism. This has particular resonance with respect to children.
13(7)	**Provided adequate welfare facilities?**			x	Schedule 2 sets out the minimum welfare facilities to be provided on a construction site. These must be provided from day one.
13(7)	**Established if welfare facilities are being effectively serviced?**			x	As well as providing welfare facilities the contractor must ensure that they remain effective.
9(2)	**Established if any work is being done which could require an amendment or modification to an existing health and safety file?**	x	x	x	A number of non-notifiable projects will inevitably involve work that could lead to a revision/update of an existing health and safety file. All duty holders must ask themselves this question.

Note:

1. All contractors must also discharge their duties under Part 4 Duties Relating to Health and Safety on Construction Sites, namely Regulations 25 to 44.
2. Contractors and clients can often take on designers' duties through:
   - procurement strategies
   - altering specifications.
3. Clients, designers and others can also take on contractors' duties through controlling the way in which construction work is carried out (Regulation 25(4)).

CHECKLIST FOR NON-NOTIFIABLE PROJECTS (Sheet 6 of 6)

10.2
CHECKLIST OF ADDITIONAL DUTIES FOR NOTIFIABLE PROJECTS

CHECKLIST OF ADDITIONAL DUTIES FOR NOTIFIABLE PROJECTS (Sheet 1 of 10)

Reg.	Action	Duty holders						Comment
		Client	Designer	Contractor	CDM co-ordinator	Principal contractor	
	Have you . . .						
14(1) and 14(5)	**Appointed a CDM co-ordinator? Such appointments must be made in writing.**	x					Appointment should be made as soon as practicable after the initial design phase. Care should be taken to appoint sooner rather than later since a late appointment would ensure the client takes on the role by virtue of the default mechanism in Regulation 14(4). The appointment should be made in writing before the appointment is taken up.
21(3)	**Signed the F10 notification**	x			(x)		Such a signature is requested from the client or someone on his behalf to endorse the fact that the client understands his duties.
14(2) and 14(5)	**Appointed a principal contractor? Such appointments must be made in writing.**	x					Whilst the timing of such an appointment would reflect the procurement strategy, such an appointment must be made before the start of any construction work. There is much benefit in appointing the principal contractor as early as possible to provide a perspective on buildability and/or constructability. Care should be taken to appoint sooner rather than later since a late appointment would ensure the client takes on the role by virtue of the default mechanism in Regulation 14(4). The appointment should be made in writing before the appointment is taken up.
14(3)	**Changed or renewed any of the above appointments?**	x			(x)		Once appointed such a function must be continuously fulfilled throughout the project. There can only be one CDM co-ordinator and one principal contractor on the project at any one time. However, the role can be transferred to another party as a result of: ● procurement strategies ● competence issues ● failure to fulfil the role. Such a change requires an amendment to the F10.
14(4)	**Noted that failure to appoint ensures that you will be deemed to function as the CDM co-ordinator and/or the principal contractor?**	x					In the context of this regulation there can no longer be any belated appointments, since failure to appoint ensures that the client takes on the full responsibilities of the CDM co-ordinator and the principal contractor.

Continued

Reg.	Action	Duty holders					Comment
		Client	Designer	Contractor	CDM co-ordinator	Principal contractor	
15	**Provided pre-construction information to the CDM co-ordinator?**	x			(x)		The provision of such information shall be made promptly. Such information should include: • information about the site • information concerning the proposed use of the structure • the minimum amount of time required for mobilisation • information in any existing health and safety file. The CDM co-ordinator must take reasonable steps to identify and collect such information.
16	**Sanctioned the start of the construction phase?**	x			(x)		This represents a critical control point to be exercised by the client dependent on the construction phase plan being developed in line with adequate welfare facilities and control methodologies for those early activities involved in construction. The client must be satisfied with the sufficiency of the construction phase plan before allowing the construction phase to start. The CDM co-ordinator is available to give suitable and sufficient advice to the client in undertaking these measures. **The duty to sanction the start of construction remains with client.**
17(1)	**Provided the CDM co-ordinator with all relevant health and safety file information in your possession?**	x					This is information which is reasonably obtainable by the client and which would be required by the CDM co-ordinator for inclusion in the health and safety file.
17(2)	**Ensured that if a single health and safety file relates to more than one project such information is easily identifiable?**	x					The format of the health and safety file should be arrived at via early discussions between the client and the CDM co-ordinator. Some clients prefer a collective approach whereby the health and safety file relates to a number of projects/sites or structures and includes other information. If this is the case care must be exercised that relevant information can be easily identified.

CHECKLIST OF ADDITIONAL DUTIES FOR NOTIFIABLE PROJECTS (Sheet 2 of 10)

Continued

CHECKLIST OF ADDITIONAL DUTIES FOR NOTIFIABLE PROJECTS (Sheet 3 of 10)

Reg.	Action	Duty holders					Comment
		Client	Designer	Contractor	CDM co-ordinator	Principal contractor	
17(3)	**Structured the management of the health and safety file so that after handover the information is:** ● **kept available for inspection by those who need it** ● **revised as often as may be appropriate?**	x					The management system must ensure the availability of the health and safety file by those who need access to it, e.g.: ● designers ● facility management ● CDM Co-ordinators whilst also ensuring the security of that document. Additionally management systems must also ensure that relevant new work/modifications/alterations provide an amendment to the health and safety file where appropriate.
17(4)	**Noted that change of ownership requires the health and safety file to be handed over to the new owner?**	x					Change of asset ownership needs to be accompanied by the transfer of any related health and safety file. The non-availability of such a document could hold up the sale transaction.
	Ensured that the transfer of the health and safety file is accompanied by an explanatory letter in respect of the nature and purpose of the health and safety file?	x					Such a transfer of the health and safety file requires an explanatory letter to ensure the new owner is aware of the purpose and nature of the health and safety file in terms of future statutory duties. A receipt for handover is prudent management.
	Have you	Client	Designer	Contractor	CDM co-ordinator	Principal contractor	
18(1)	**Ensured a CDM co-ordinator has been appointed?**		x				Design should not progress beyond the initial design phase until such an appointment has been made. To do so would expose the client to take on the role of CDM co-ordinator by virtue of the default mechanism associated with Regulation 14(4).
18(2)	**Provided sufficient information about relevant aspects of the design of the structure, its construction and/or its maintenance as will enable the CDM co-ordinator to discharge his duties?**		x				The design process must communicate relevant information to the CDM co-ordinator. Such information would be relevant for inclusion in pre-construction information to be forwarded to other designers and contractors who have been or who might be appointed by the client. Similarly such information would also be relevant for the health and safety file.

Continued

Reg.	Action	Duty holders					Comment
		Client	Designer	Contractor	CDM co-ordinator	Principal contractor	
19(1)(a) (b) (c)	**Been provided the names of the CDM co-ordinator and principal contractor?**						**No construction work should start until:** (a) such names have been received
	Been given access to the relevant parts of the construction phase plan?			x			(b) each contractor is in receipt of detailed relevant information; this is critical to the suitable and sufficient control of the work interfaces
	Ensured notice of the project has been given to the HSE or the Office of Rail Regulation?			x			(c) notice has gone off to the relevant office. Items (a) and (b) can be gleaned from the display of the F10 notification, which should be in a prominent position on site.
19(2)(a)	**Promptly provided the principal contractor with information which might:** • **affect the health and safety of any person carrying out or affected by construction** • **justify a review of the construction phase plan** • **be included in the health and safety file?**			x			Site management demands a proactive approach in a **prompt** manner. Constant vigilance and feedback up the chain of management is required throughout all phases of construction.
19(2)(b)	**Identified to the principal contractor all contractors appointed or engaged by you?**			x			Such identification needs to done **promptly** and enables the principal contractor to effectively manage all relevant aspects of the project through his knowledge and awareness of all parties on his site.
19(2)(c)	**Complied with:** • **directions given by the principal contractor** • **any site rules?**			x			Such directions must be reasonable and are there to guide and protect individuals at all times.
19(2)(d)	**Provided relevant information to the principal contractor for the purpose of notification or report under RIDDOR 1995?**			x			As well as providing information to the principal contractor, the duty holder must also file the notification or report to the HSE.
19(3)(a)	**Taken reasonable steps to ensure that construction work is being carried out in accordance with the construction phase plan?**			x			Such steps include the planning, managing and monitoring of construction work, so ensuring that it is carried out without risks to health and safety.

CHECKLIST OF ADDITIONAL DUTIES FOR NOTIFIABLE PROJECTS (Sheet 4 of 10)

Continued

CHECKLIST OF ADDITIONAL DUTIES FOR NOTIFIABLE PROJECTS (Sheet 5 of 10)

Reg.	Action	Duty holders					Comment
		Client	Designer	Contractor	CDM co-ordinator	Principal contractor	
19(3)(b)	**Taken appropriate action to ensure health and safety where the construction phase plan is found to be deficient?**			x			This should be a short-term action response until the construction phase plan has been modified by the principal contractor.
19(3)(c)	**Notified the principal contractor of significant findings requiring an alteration of the construction phase plan?**			x			Emphasis is placed on the proactive approach of all duty holders. If systems are not working then feedback needs to be made to address the shortcomings.
20(1)(a)	**Given suitable and sufficient advice and assistance to the client to enable him to effectively discharge his duties?**				x		The CDM co-ordinator needs to work closely with the client's team throughout but particularly at the start of any project. Particular reference is made to assistance in complying with management arrangements (Regulation 9) and control over the start of construction (Regulation 16).
20(1)(b)	**Ensured that suitable arrangements have been made and implemented for the co-ordination of health and safety measures during planning and preparation for the construction phase, including:** • **co-operation/co-ordination** • **application of the general principles of prevention?**				x		The CDM co-ordination team must be integrated within the project team to be assured that suitable arrangements are made and implemented for general co-ordination and the facilitation of co-operation (Regulation 5), co-ordination (Regulation 6) and that general principles of prevention are applied (Regulation 7). This requires liaison with all duty holder groups.
20(1)(c)	**Liaised with the principal contractor regarding:** • **the contents of the health and safety file** • **information required by the principal contractor for the construction phase plan** • **design developments which may affect planning and management?**		(x)		x	(x)	As a result of liaison the link with the principal contractor is reinforced throughout the construction phase.

Continued

Reg.	Action	Duty holders					Comment
		Client	Designer	Contractor	CDM co-ordinator	Principal contractor	
20(2)(a)	**Identified and collected pre-construction information?**	(x)			x		Reasonable steps must be taken to identify and collect relevant information from client and designers. This should be done through dialogue and consultation with the assembled professional teams and not be a unilateral course of action taken by the CDM co-ordinator unless absolutely necessary.
20(2)(b)	**Promptly provided relevant pre-construction information to:** ● **all designers** ● **every contractor appointed or who might be appointed by the client?**		(x)	(x)	x		Pre-construction information must be provided to all designers and contractors who have been or who may be appointed by the client. Such information could be provided well into the construction phase depending on the procurement strategy.
20(2)(c)	**Taken all reasonable steps to ensure that designers are complying with their duties under Regulations 11 and 18(2)?**				x		This is best facilitated through invited attendance at review meetings and through intercourse at development and progress meetings.
20(2)(d)	**Taken all reasonable steps to ensure co-operation between designers and the principal contractor in relation to any design or change to a design?**				x		This focuses on the critical design development or change during the construction phase, where the principal contractor is now restrained by the confines of the construction programme. Good communication between the design process and construction is essential to address related implications on construction control. Particularly relevant is the liaison demanded between permanent and temporary design.
20(2)(e)	**Prepared and/or updated the health and safety file?**				x		**Note:** It is now the CDM co-ordinator's legal duty to prepare the health and safety file. This does not translate into the CDM co-ordinator having to actually prepare the health and safety file, for arrangements can be made for others to prepare it, but it remains the CDM co-ordinator's duty to ensure the management of the preparation process.
20(2)(f)	**Passed the health and safety file to the client?**	(x)			x		Timing is critical and the handover should coincide with the end of the construction phase. Similarly, if there are phased handovers or sectional completions the health and safety file handover should be structured to reflect these arrangements.

Continued

CHECKLIST OF ADDITIONAL DUTIES FOR NOTIFIABLE PROJECTS *(Sheet 6 of 10)*

CHECKLIST OF ADDITIONAL DUTIES FOR NOTIFIABLE PROJECTS (Sheet 7 of 10)

Reg.	Action	Duty holders						Comment
		Client	Designer	Contractor	CDM co-ordinator	Principal contractor		
	Ensured an explanatory letter has accompanied the health and safety file?				x			Such a letter satisfactorily concludes the handover exercise and should explain to the client what future liabilities are incurred in his ownership of the health and safety file (Regulation 17(3)).
	Received a signature for the receipt of the health and safety file?	(x)			x			It would be imprudent to hand over the health and safety file without receiving a signature for its receipt from the client or someone representing the client. This will avoid any future arguments.
21(1)	**Sent notification off to the relevant authority?**				x			Such a notification should have gone either to the HSE or the Office of Rail Regulation depending on the nature of the work being undertaken.
21(2)	**Forwarded additional information?**				x			This relates to the usual process of a two-stage notification: initially on the appointment of the CDM co-ordinator and then additional information at the appropriate time, as soon as practicable after the appointment of the principal contractor. Should relevant information change during the project then an amendment should be forwarded by the CDM co-ordinator.
21(3)	**Ensured the client or someone on his behalf has signed the notification?**	(x)			x			It is important that the client, or someone strategically placed representing the client, signs to signify they are aware of their duties under these regulations. Electronic signatures are acceptable if their use is approved by the client.
	Have you . . .	Client	Designer	Contractor	CDM co-ordinator	Principal contractor		
22(1)(a)	**Planned, managed and monitored the construction phase?**					x		Any planning and associated management is only as good as the accompanying monitoring regime. This process must continue throughout the construction phase facilitated by all duty holders' roles in co-operation (Regulation 5), co-ordination (Regulation 6) and the application of general principles of prevention (Regulation 7).

Continued

Reg.	Action	Duty holders					Comment
		Client	Designer	Contractor	CDM co-ordinator	Principal contractor	
22(1)(b)	**Liaised with the CDM Co-ordinator?**		(x)		(x)	x	Liaison is essential for the co-operation between designers and the principal contractor in matters of response to ongoing design and/or changes in design.
22(1)(c)	**Ensured welfare facilities are sufficient?**					x	Compliance with the requirements of Schedule 2 must begin from day one of the construction phase.
22(1)(d)	**Drawn up rules?**					x	Specific site rules for health and safety management are necessary to safeguard all those on site. Such rules should be: ● in writing ● understandable ● brought to everyone's attention ● enforced ● displayed in a prominent position on site ● in comprehensible language and reflect the ethnic mix on site.
22(1)(e)	**Given reasonable directions?**					x	Reasonable directions must be given to any contractor to enable the principal contractor to fulfil his duties.
22(1)(f)	**Notified every contractor of the minimum mobilisation period?**			(x)		x	This is the period for planning and preparation before that specific item of construction activity starts. Ideally this should be ratified by dialogue between contractor and principal contractor.
22(1)(g)	**Consulted contractors before finalising relevant parts of the construction phase plan?**			(x)		x	Consensual ownership of health and safety is promoted by consultation, particularly in respect of areas of common interest and involvement.
22(1)(h)	**Given access to every contractor to the construction phase plan?**					x	Timely access to the relevant parts of the construction phase plan is necessary to appreciate the full implications of the related work.
22(1)(i)	**Provided further information to every contractor?**			(x)		x	Timely provision of information before the contractor begins work is needed to enable full compliance with welfare provision (Schedule 2) and to ensure work is without risk, so far as is reasonably practicable, to the health and safety of any person.
22(1)(j)	**Identified each contractor's input to the health and safety file?**			(x)	(x)	x	Early identification of inputs facilitates the management of the health and safety file by the CDM co-ordinator. As well as inputs the time for delivery also needs to be stipulated. Such inputs should initially be identified from dialogue between the principal contractor and the CDM co-ordinator.

Continued

CHECKLIST OF ADDITIONAL DUTIES FOR NOTIFIABLE PROJECTS (Sheet 8 of 10)

CHECKLIST OF ADDITIONAL DUTIES FOR NOTIFIABLE PROJECTS (Sheet 9 of 10)

Reg.	Action	Duty holders					Comment
		Client	Designer	Contractor	CDM co-ordinator	Principal contractor	
22(1)(k)	**Displayed the F10 notification?**					x	The notification to the HSE or Office of Rail Regulation should be displayed in a readable condition in a position where it can be read by any worker engaged in construction work. Such a display needs to be continuous.
22(1)(l)	**Taken reasonable steps to secure the site?**					x	The principal contractor is responsible for securing the boundaries of the site throughout the construction period. Reasonable steps must be taken. **Note:** Site boundaries often change depending on phasing and handover arrangements. ACoP notes that special consideration is needed for: ● rights of way ● interfaces with other work areas ● partial handovers, particularly on housing sites ● children and other vulnerable people.
22(2)	**Provided:** ● **site induction** ● **information and training** ● **any further information and training?**			(x)		x	Collectively this represents a vital component of communication by site management in respect of health and safety. Such information can be cascaded down the supply chain by others in charge after they have received it. An update of such information must be given that is compatible with changing phases on the construction site. An induction site log is generally used to indicate receipt of such information.
23(1)(a) and 23(1)(b)	**Ensured the construction phase plan is:** ● **developed sufficiently before starting construction work** ● **developed thereafter?**	(x)			x	x	The principal contractor is charged with developing the construction phase plan sufficiently before the construction phase starts. Adequate attention must be paid to: ● welfare provision ● workplace risk assessments and control measures ● management arrangements, etc. all as outlined in Appendix 3. **NO CONSTRUCTION WORK COMMENCES UNTIL THE CLIENT HAS SIGNIFIED HIS SATISFACTION WITH THESE ARRANGEMENTS.**

Continued

Reg.	Action	Duty holders					Comment
		Client	Designer	Contractor	CDM co-ordinator	Principal contractor	
23(1)(c)	**Arranged for construction work to be implemented safely?**					x	Implementation of the construction phase plan must ensure, so far as is reasonably practicable, the health and safety of all those carrying out construction work or affected by such work. It therefore embraces the health and safety of many others besides construction workers, including members of the public, visitors and trespassers.
23(2)	**Identified the risks in the construction phase plan and included appropriate control measures?**					x	The identification of risks and the suitable and sufficient measures to manage and control such risks are the domain of the principal contractor. This function is aided and abetted by all contractors on site. If the identification process is flawed, controls thereafter are limited.
24	**Co-operated and consulted with workers?**					x	Co-operation and consultation with the workforce and/or their representatives is a cornerstone of health and safety ownership, which should be achieved preferably by consensus and not imposition. This relationship benefits from monitoring, feedback and amendments to arrangements and controls. Inspection and copying of relevant information is part of this process. Timely consultation is essential. Exceptions occur (refer to ACoP, page 80).

Note: All work on a construction site must comply with Part 4 Duties Relating to Health and Safety on Construction Sites, namely Regulations 25 to 44 , and other related legislation as appropriate.

CHECKLIST OF ADDITIONAL DUTIES FOR NOTIFIABLE PROJECTS (Sheet 10 of 10)

10.3
INITIAL DESIGN/CLIENT MEETING: PROMPT LIST

No.	Reg.	Description	Yes	No	Comments
1		Client identity?			Established? Elected client?
2	3	Non-notifiable or notifiable			Established?
3	11(1)	Client awareness of his duties?			Communicated? How?
4		Procurement strategy			Established?
5	12	Design undertaken outside Great Britain?			Commissioning link identified?

6	9(1)	Start date for construction:
		• planning approval
		• design development
		  – feasibility
		  – conceptual design
		  – detailed design
					Established and suitable?

7		Initial design/significant detailed design			Establish demarcation
8	14(1)	CDM co-ordinator			Appointed?
9	9(1)	Construction time			Suitable?
10	9(1)	Resources			Suitable?
11		Appointed parties			Identify other designers? Identify contractors?

12	10	Information:
		• health and safety file
		• asbestos management surveys
		• other surveys:
		  – services
		  – condition
		  – geotechnical
		  – ground
		• any other relevant information about the site
		• other sources
		• programme and key dates
					Available?
					Required?
					
					Established?

13	10(2)	Use of the building			Established?
14	10(2)	Mobilisation period			Established?
15		Client's management arrangements			Suitable?
16	5	Co-operation			Other parties identified?
17	6	Co-ordination			Other parties identified?
18	ACoP paragraph 20	Lead designer?			Established?

10.4
PRE-START MEETING: PROMPT LIST (NON-NOTIFIABLE PROJECTS)

No.	Reg.	Description	Yes	No	Comments
All projects		Determine whether the project is non-notifiable or notifiable			
Introduction					
1.	2	Identification of duty holders: • client • designers • contractors			Names, points of contact, telephone numbers, e-mails, etc.
Co-operation					
2.	5(1)(a)	Identification of parties on this project			Names, points of contact, telephone numbers, e-mails
3.	5(1)(b)	Identification of parties on adjoining sites			Names, points of contact, telephone numbers, e-mails
Co-ordination					
4.	6	Identification			Names, points of contact, telephone numbers, e-mails
5.	ACoP paragraph 48	Lead designer			Established?
6.	ACoP paragraph 20	Main contractor			Established?
Arrangements					
7.	9(1)	Management arrangements			Suitability?
8.	9(1)	Construction period			Verification by contractor(s)?
9.	9(1)	Roles and responsibilities			Established?
10.		Communication between parties			Protocols established: • links • points of contact • frequency of meetings etc.
Information					
11.	10(2)	Information			Awaited? Received?
12.	10(1)	Pre-construction information after the construction start			Identify any designers to be appointed after the start of construction and contractors directly appointed by client
13.	10(2)(c)	Mobilisation period			Ratified by contractor(s)
14.		Health and safety file			Amendments required to existing?
Design process					
15.	12	Design undertaken outside Great Britain?			Identified? Establish commissioning link
16.	11(6)	Provision of information after the start of construction: • design development • design change			Protocols established?
Contractors					
17.	13(3)	Mobilisation period			Ratified by contractor?
18.	13(4)	Training			Information and training provided?
19.	ACoP paragraph 52	Protection			Workers and public?
20.	13(6)	Security of the site			Established?
21.	13(7)	Welfare provision			Compatible with Schedule 2 from day one?

10.5
PRE-START MEETINGS: PROMPT LIST (NOTIFIABLE PROJECTS) AND ADDITIONAL POINTS

No.	Reg.	Description	Yes	No	Comments
Additional points for notifiable projects					
Introduction					
A.	14(1)	Identification of duty holders: • CDM co-ordinator • principal contractor			Names, points of contact, telephone numbers, e-mails, etc.
B.	14(5)	Appointments in writing			Principal contractor? (CDM co-ordinator should already have been appointed after the initial design phase)
Notification					
C.	Schedule 1 21(2)	Additional F10 to be forwarded			Copy handed to principal contractor for display on site? (Initial F10 should have been signed by client or someone on his behalf)
Construction start					
D.	16 and 23(1)(a)	Sufficiency of the construction phase plan			Established? (Adequate welfare facilities on site from first day)
E.	23(1)(a)	Adequate methodology with respect to safe and suitable systems of work			Critical early stage site activities appropriately articulated? (Later activities can be articulated as part of the development of the construction phase plan)
F.	16	Sanction to start construction			Received? (Client must control the start of construction depending on the construction phase plan being sufficiently developed. CDM co-ordinator provides advice and assistance)
Health and safety file					
G.	22(1)(j)	Input from contractors			To be identified

10.6
PROGRESS MEETING: PROMPT LIST

No.	Reg.	Description	Yes	No	Comments
Notifiable projects					
	Notification				
1	22(1)(j)	F10 on display			In a prominent position?
	Construction process management				
2	23(1)(b)	Construction phase plan development			Compatibility with the programme of works?
3	23(1)(b) and 23(1)(c)	Construction phase plan development			Contractors' review of effectiveness? Action taken?
4	22(1)(b)	Design development change Design amendment			Construction phase plan compatibility?
5		Inspection reports and site safety audits			Adverse comments? Action to be taken? RIDDOR reports?
6	Appendix 2	Health and safety goals			Achieved or not achieved? Action required?
7		HSE			Any visits? Outcomes to visit?
8	13(4) and 22(2)	Site induction log			Up to date?
9		Re-induction requirements			Updates?
10	13(6) and 13(1)(l)	Site security			Security breaches? Boundary changes?
11	13(7) and 22(1)(c)	Welfare			Effective? Compliant with Schedule 2?
12	22(1)(f)	Mobilisation			All contractors informed?
13	22(1)(j)	Health and safety file			Inputs identified? Receipt of information?
	Co-operation				
14	5	Additional parties for co-operation			Overlapping projects? Identify?
	Co-ordination				
15	6	Additional parties for co-ordination			Identify?
	Arrangements				
16	9	Management arrangements: ● CDM co-ordinator ● designer(s) ● principal contractor ● contractor(s)			Effective?

Section 11
COMPETENCE

The Health and Safety Executive's Research Report RR422 *Competency and Resource*, edited by John Carpenter, provided a discussion document which raised the topic of competence in the discharge of duties, with particular reference to the CDM Regulations.

Many of the points raised by John Carpenter have now been embraced by the CDM Regulations 2007 and Appendices 4 and 5 of the Approved Code of Practice, together with an accompanying section focusing directly on competence and training in a manner that previous documentation has not addressed.

Direction is provided for all appointed duty holders, i.e. CDM co-ordinators, designers, principal contractors and contractors together with assessments in respect of the competence of organisations.

The competence and adequacy of resourcing have always been cornerstones of appointment and yet a superficial approach has been adopted by some, which undermines the entire credibility of the appointment process. The construction project is hostile territory and is no place for the maverick, whatever position he occupies. Hopefully the self-certification of competence should emphasise this perspective and ensure that there is a moral contract between those who appoint and those who accept.

As noted in paragraph 194 of the ACoP:

'Assessments should focus on the needs of the particular project and be proportionate to the risks, size and complexity of the work.'

Competence assessment is based on task knowledge, experience and ability, as well as acknowledging that it must be a continuing business and dependent on continuing professional development to update knowledge and maintain competence at a threshold acceptable to the industry and the legislation. It has a new resonance in that all duty holders in accepting appointments now self-certify that they are competent (Regulation 4(b)) to perform the function.

The assessment of competence must have a rigour compatible with the task and the complexity of the project itself.

The ACoP suggests that competency assessments of organisations should be carried out against core criteria as a two-stage process:

203. Stage 1:

An assessment of the company's organisation and arrangements for health and safety to determine whether these are sufficient to enable them to carry out the work safely and without risk to health.

204. Stage 2:

An assessment of the company's experience and track record to establish that it is capable of doing the work; it recognises its limitations and how these should be overcome and it appreciates the risks from doing the work and how these should be tackled.'

Such assessments should be made against those criteria outlined in Appendix 4 of the ACoP.

A similar two-stage process is advocated for individuals based on factors of individual qualifications and training records together with continuing professional development as well as past experience in work of the nature required by the project.

Regulation 4(1)(c) also states that no one shall:

'arrange for or instruct a worker to carry out or manage design or construction unless the worker is:

(i)   *competent, or*
(ii)   *under the supervision of a competent person.'*

It is interesting to note that whilst the ACoP provides examples of evidence by which the achievement of a suitable standard can be demonstrated, the construction market place is already using such standards as criteria for appointment. Whilst such an endorsement is useful it could also be limiting, particularly in consideration of the wide catchment area occupied by construction professionals.

Registers are mentioned in the ACoP as being a factor in competence assessment and in particular those operated for designers and CDM co-ordinators by the Institution of Civil Engineers, the Association for Project Safety and the Institution of Construction Safety. The Association for Project Safety requires individuals wishing to be placed on the register for CDM co-ordinators to achieve a minimum of 18 points. These can be amassed from professional qualifications, health and safety qualifications, and experience, with a minimum of two points obtained in the health and safety area and a minimum of two points obtained in the design area. The Health and Safety Register operated by the Institution of Civil Engineers is only open to its members. The Institution of Construction Safety was unable to provide any further information at the time of writing. Additionally, membership of a relevant professional institution has obvious benefits for those operating as designers.

CDM co-ordinators certainly need a working appreciation of the design process, as well as the necessary insight to the construction process, supplemented by a level of health and safety awareness generally and supported by specific knowledge of the CDM Regulations 2007. Experience of similar projects should reflect the complexity of the project in question. The skill-set of the obsolete planning supervisor is a useful platform but obviously needs to be extended, as articulated in Appendices 4 and 5 of the ACoP.

The liaison and communication demanded of the CDM co-ordinator through interface with the other duty holders ensures that heavy reliance is placed on good interpersonal skills.

Training schemes and 'toolbox' talks figure heavily in ensuring site operatives also follow a programme of training to raise levels of competence to those appropriate to the modern demands of the construction site.

For design, construction and CDM co-ordination advocated on the more complex project, a team-based approach will inevitably be sought and a corporate appointment made. There is much sense in the organisation seeking to achieve corporate competence, as advocated in Appendix 4 of the ACoP for this very purpose and demonstrated through an organisational network that is structured to give visibility to the empowerment of individuals within the supervisory context of the competent person.

It is only through the raising of competence levels across all sectors of the construction industry, coupled with a greater management commitment, that any legislation will achieve the cultural change demanded to challenge the unacceptable level of fatalities and occupational ill-health and disease endemic in the construction industry.

WEB PAGE DIRECTORY

No.	Page	Description
1	www.aps.org.uk	The Association for Project Safety
2	www.bre.co.uk	Building Research Establishment
3	www.cibse.org	Chartered Institute of Building Services Engineers
4	www.thecc.org.uk	Construction Confederation
5	www.constructingexcellence.org.uk	Constructing Excellence
6	www.cic.org.uk	Construction Industry Council
7	www.citb.org.uk	Construction Industry Training Board
8	www.cpa.uk	Construction Plant Hire Association
9	www.dti.org.uk	Department of Trade and Industry
10	www.agency.osha.eu.int	European Agency for Safety and Health at Work
11	www.hse.gov.uk	Health and Safety Executive: United Kingdom
12	www.hsebooks.com	Health and Safety Executive Books
13	www.hsa.ie	Health and Safety Authority : Ireland
14	www.ice.org.uk	The Institution of Civil Engineers
15	www.imeche.org.uk	The Institution of Mechanical Engineers
16	www.istructe.org.uk	The Institution of Structural Engineers
17	www.nao.org.uk	National Audit Office
18	www.demolition-nfdc.com	National Confederation of Demolition Contractors
19	www.nhbc.co.uk	National House-Building Council
20	www.ogc.gov.uk	Office of Government Commerce
21	www.eustatistics.gov.uk	Office for National Statistics
22	www.pff.org.uk	Precast Flooring Federation
23	www.rias.org.uk	The Royal Incorporation of Architects in Scotland
24	www.architecture.com	Royal Institute of British Architects

Continued

No.	Page	Description
25	www.rics.org	The Royal Institution of Chartered Surveyors
26	www.rospa.com	The Royal Society for the Prevention of Accidents
27	www.safetyindesign.org	Safety in Design Ltd
28	www.scoss.org.uk	Standing Committee on Structural Safety
29	www.strategicforum.org.uk	Strategic Forum
30	www.design4health.com	University of Loughborough

Bibliography

CDM Procedures Manual, 2nd edn, S. D. Summerhayes, Blackwell Publishing, Oxford, 2002.

CDM 2007 Construction work sector guidance for designers (C662), O Arup and Partners, CIRIA, London, 2007.

Construction (Design and Management) Regulations 2007. Approved Code of Practice, Managing Health and Safety in Construction (L144), HSE Books, 2007.

Construction (Design and Management) Regulations 2007, HMSO Statutory Instrument 2007 No. 320, The Stationery Office Limited, London, 2007.

Five Steps to Risk Assessment, INDG 163 (rev. 1), HSE Books, 1998.

HSE RR422, *Competency and Resource*, J. Z. Carpenter, HSE Books, 2006.

Management of Health and Safety at Work Regulations 1999. Approved Code of Practice and Guidance L121, 2nd edn, HSE Books, 2000.

Newsletters (published approximately every two months), The Association for Project Safety, 12 Stanhope Place, Edinburgh EH12 5HH.

Practice Notes (published approximately every two months), The Association for Project Safety, 12 Stanhope Place, Edinburgh EH12 5HH.

Research Report 218, Peer Review of Analysis of Specialist Group Reports on Causes of Construction Accidents, HSE Books, 2004.

Research Report 422, Developing Guidelines for the Selection of Designers and Contractors under the Construction (Design and Management) Regulations 1994, HSE Books, 2005.

Rethinking Construction, the Report of the Construction Taskforce, chaired by Sir John Egan, Department of the Environment, Transport and the Regions, London, 1998.

Revitalising Health and Safety in Construction, Discussion Document, HSE Books, 2002.

Successful Health and Safety Management, HSG 65, 2nd edn, HSE Books, 1997.

Workplace Health, Safety and Welfare. Workplace (Health, Safety and Welfare) Regulations 1992. Approved Code of Practice L24, HSE Books, 1992.

INDEX

Please also see the Contents for a comprehensive list of subjects covered in this book